QED COHERENCE
IN MATTER

QED COHERENCE IN MATTER

Giuliano Preparata
Università Statale di Milano

World Scientific
Singapore • New Jersey • London • Hong Kong

Published by

World Scientific Publishing Co. Pte. Ltd.

P O Box 128, Farrer Road, Singapore 9128

USA office: Suite 1B, 1060 Main Street, River Edge, NJ 07661

UK office: 57 Shelton Street, Covent Garden, London WC2H 9HE

Library of Congress Cataloging-in-Publication Data

Preparata, Giuliano.
    QED coherence in matter / Giuliano Preparata.
        p.   cm.
    Includes bibliographical references.
    ISBN 9810222491
    1. Condensed matter.   2. Quantum electrodynamics.   I. Title.
QC173.454.P74   1995
537.6'7--dc20   95-13463
                                                    CIP

Cover illustration published with permission of the Ministero per i Beni Culturali ed Ambientali.

Printed in Singapore.

To my father Vincenzo,
who taught me the meaning
of honour and honesty.

FOREWORD

I knew that, eventually, I had to write a book: my colleagues, my pupils, many acquaintances have been asking me time and again to do something to bring the research work of the last years (roughly from 1988 onwards) on Quantum ElectroDynamics (QED) in matter closer to them, in the form of an orderly presentation of the ideas and of the results which, so far, have only seen the difficult and dim light of the learned journals.

The unfortunate occasion of a bad fall, causing the fracture of a wrist (fortunately the left one), in the summer of 1992 appeared to me as an act of God, which was unmistakably summoning me to my duty. I took it with good grace and plunged with a fury, that quite surprised me, into the deep waters of the physics of condensed matter, in which I have been only a neophyte, my trade having been for almost thirty years theoretical High Energy Physics. The point is that I am still a neophyte, the subject being so unfathomably immense that even those who have devoted their whole life to condensed matter physics have mostly become "experts" of well defined and restricted fields. So, how could I dare to undertake so lightly such a task, incommensurate to the little knowledge that I had acquired and to the endeavor large but inadequate that I had put into the enterprise? But the summon sounded so loud and clear that I had no choice: I had to carry on. And, after almost two years, here it is.

This is not a textbook, but rather a book of texts. The eleven Chapters in which it is subdivided mostly expand and, sometimes, deepen the material presented in the papers and lectures that I have been publishing since 1988 on the uncertain steps of a new approach to condensed matter, based on a full application of QED. It would seem surprising that after so many years of QED (the foundation of QED is only slightly posterior to that of Quantum Mechanics) its potentialities have not been thoroughly understood and exploited. But, as I discuss in Chapter 3, the great successes of Laser Physics (LP), where the subtle order brought about by the coherent interaction between the electromagnetic field and the atomic systems shines in full glory, paradoxically and ironically have grown into a sort of psychological barrier towards recognizing its presence in the quantum-mechanical ground states of condensed matter. The problem has been that in LP, being built upon excited states, the basic QED equations can be accurately described by means of an approximation — the "slowly varying envelope" approximation — that predicts coherent solutions for the matter and e.m. systems **only** in excited, unstable states, which for their sustainment thus require very special external conditions: an external power source — the "pump" — and a finely tuned set of mirrors — the "cavity" —. No "pump", no "cavity" — Anderson's "dictum" dictates[1] — no laser! And the inescapable consequence is that the ground state of any piece of condensed matter must be built upon the free-field (perturbative) ground states of the e.m. fields as well as of the matter fields: the only interactions allowed are the various forms of electrostatic interactions, which, due to the electric neutrality of matter at spatial scales of the order of 10^{-8} cm, have thus such a very short range. How devilishly difficult it is to build long-range ordered configurations with short-range interactions is clearly demonstrated by the fact that we still do not have a satisfactory theory of magnetism, superconductivity both cold and hot, superfluidity of both ^{4}He and ^{3}He, not to mention the strange behaviours of Pd-deuterides and of water itself.

Wouldn't, perhaps, all of it change if Anderson's "dictum" were just wrong? Couldn't one understand the deep reasons for the marvellous order one observes in matter, if the ordered states, familiar in LP, could occur in the ground state as well? These were the questions that I had in my mind, when in 1987 I set up to "have a try" at QED in condensed matter. I had just completed perhaps the most difficult calculation of my whole life as a theoretical physicist: the variational determination of the "effective potential" of a non-abelian gauge theory, such as Quantum ChromoDynamics (QCD), in an external "chromomagnetic" field[2]. To my big surprise a typical non-perturbative (variational) calculation showed that even for very small coupling constants g, the variational ground state was very far away from the perturbative ground state (PGS), that differs from the free-field theory's one by $O(g^2)$ (perturbative) corrections. This was indeed very surprising, for it meant the inadequacy of the generally accepted ideas about QCD, which turn around the important notion of Asymptotic Freedom (AF), that since its discovery in 1973 has become a kind of "taboo", something that cannot be wrong. Nevertheless this strange phenomenon, that would resist all my attacks to make it disappear, to my even bigger surprise was sort of deleted from the conscience of the community of High Energy Physics, and with it its discoverer. A very painful and unexpected fact to contemplate for a convinced follower of the galilean tradition. But, God be thanked, the joys have by far exceeded the sorrows, for out of that surprising result the first viable theory of colour confinement has emerged, and with it a well defined new perturbation theory[3] that allows one to calculate all possible aspects of the physics of hadrons to the desired accuracy. And if the QCD ground state was so highly non-perturbative, would'n it be reasonable to expect a similar situation to hold for the QED ground state in condensed matter? After all the electric dipoles (because of matter neutrality) present everywhere in condensed matter could somehow mimic the dynamics of non-abelian gauge theories, one of the gauge fields being the e.m. field and the charged fields being realized by the waves of electric polarization.

Once this intuition was given an adequate quantum field theoretical formulation[4], all the rest easily followed, the struggle being only against one's own ignorance of the wonderfully immense world of condensed matter physics. In this way, in the thermodynamic limit when the temperature and the density were right from the QED field equations there emerged **spontaneously** a world of order and harmony that, I thought, I was given the great chance to be the first to contemplate. But I felt no pain when, two years later, I discovered that two eminent mathematical physicists Klaus Hepp and Elliot Lieb[5] had found all this fifteen years earlier, while studying the statistical thermodynamics of the Dicke's Hamiltonian, a simple model of atomic systems coupled to radiation which was proposed in 1953[6] by the true pioneer of QED coherence in matter, R.H.Dicke. The question then was: why was all this forgotten? Why did condensed matter physicists let go such a beautiful mechanism of long range order? The answer is that soon after the fundamental work of Hepp and Lieb, a numbers of papers remarked that Dicke's Hamiltonian was only a model which "unfortunately" violated explicitly the requirements of "gauge invariance": should one correct this flaw the results of Hepp and Lieb would evaporate and no transition to order — the Superradiant Phase Transition (SPT) — would occur in the Dicke's system of N two-level atoms.

When I learned about all this I was already deeply involved in this research programme, I thus felt the desperation that accompanies the end of a beautiful dream. But, would it be just a dream? Could a research program that had already proven its vitality be destroyed by such a simple formal flaw? I began to doubt it: now we know[7] that Hepp and Lieb were right, and that their critics had made a trivial, rather unpardonable mistake. But why didn't Hepp and Lieb and all the condensed matter physics community react more vigorously? To this day I have found no reasonable answer to this question, which, perhaps, regards sociology more than science: however, "hypotheses non fingo".

It is thus a pleasant circumstance for me to end two years of labour writing this Foreword and see all the perils that, however, could not sink this enterprise. As I emphasize time and again this book is not meant to be complete and finished like a textbook, if that were my aim this book would not see the light for many years to come. Its subject matter is in such a state of instability and quick change, that I should, no doubt, already be rewriting its eleven Chapters for another couple of years of hard work. But let it be! There is always time for revisions: the message it carries is now for me, by far, the most important aim. Let the reader forgive me for its difficult reading and the mistakes that have still escaped detection.

Finally let me thank wholeheartedly my collaborators and pupils, without whom this book would not exist: in particular Emilio Del Giudice, Renata Mele, Massimo Scorletti, Marco Verpelli, Phil Ratcliffe, She-Sheng Xue, Ruggero Barni, Daniele Garbelli, Stefano Sanvito, Fabio Taddei, Stefano Villa, Luca Gamberale, Marina Gibilisco, Luca Pirovano, Renzo Alzetta and Giuseppe Liberti. A special thank should also go to Louis Dick and Charles P. Enz for their friendship and support.

Milano 22 March, 1994.

References

1. P. W. Anderson, *Basic Notions of Condensed Matter Physics* (Benjamin-Cummings, Menlo Park (Ca), 1984)
2. G. Preparata, *Il Nuovo Cimento* **96A** (1986) 366.
3. G. Preparata, *Il Nuovo Cimento* **103A** (1990) 1073.
4. G. Preparata, *Phys. Rev.* **A38** (1988) 233.
5. K. Hepp and E. Lieb, *Ann. Phys.* **76** (1973) 360; *Phys. Rev.* **A8** (1973) 2517.
6. R.H. Dicke, *Phys. Rev.* **93** (1954) 99.
7. E. Del Giudice, R. Mele and G. Preparata, *Mod. Phys. Lett.* **B7** (1993) 1851.

CONTENTS

Chapter 1

QUANTUM FIELD THEORY OF MATTER AND RADIATION

1.1 Quantum fields: the collective description of very large ensembles of elementary quantum systems

Classical physics, the powerful view of natural phenomena that dominated modern science since its emergence from Renaissance thought right up to the beginning of the twentieth century, can well be defined as that picture of Nature that most conforms to our sensory experience. Even though it took more than two millennia to liberate natural philosophy from some obvious delusions arising from the immediate sensory perceptions (and to think of the law of the inertia, as it first occurred to Galileo Galilei), classical physics is deeply characterised by its unwillingness to "feign hypotheses" that go much beyond the sensory data, and its demand of "concrete" (mechanical) models to explain any given class of phenomena.

It is thus that in classical physics continuity came to be intimately associated with the **field** concept, while discontinuity found its root in the notion of **particle**, the irreducible quantum of matter and energy. In this way the world of natural phenomena could be sharply separated into two broad domains: wave-phenomena, on one hand, typical of the dynamical behaviour of fields, and particle phenomena, involving the interaction of matter points following the all-encompassing Newtonian paradigm, on the other. In spite of the analogies and relationships discovered in the dynamical descriptions of fields and particles by intellectual giants such as Euler, Lagrange and Laplace, and the "unification" efforts that led to the Hamilton-Jacobi theory, waves and particles remained irreducible, the former being associated with continuous and the latter with discontinuous phenomena.

As we know, this fundamental dichotomy, which is so vividly intuitive, is at the root of all "interpretational" difficulties of Quantum Mechanics (QM), that found it necessary to do away with it, by attaching to Newton's matter point a wave, the de Broglie-Schrödinger wave, so as to explain, for instance, the strange phenomena of electron diffraction and of the quantization of atomic levels.

The hard realities of atomic phenomena thus forced us to go beyond Classical Mechanics (CM), and its world of intuitive certainties, by abandoning the wave-particle dichotomy and with it the notion of well defined particle trajectories, that is quantitatively disposed of by the Heisenberg's indeterminacy relations. However, in the first big step beyond CM, the QM of a finite set of particles, the so-called "first-quantized theory", though the notion of matter point was so to speak "drowned in the sea" of its continuous wave-function, it nevertheless retained some deep, irreducible elements of discontinuity: the so-called "wave-particle duality", or "complementarity" to adopt Bohr's definitions. Indeed even though the (complex) wave-field associated with a quantum-mechanical matter point enjoys all properties of continuity, differentiability etc. that a classical wave-field is engendered with, nevertheless its phenomenology is **not** that of a classical wave-field, like for instance a classical charged fluid, for our observations on, say, a beam of electrons do reveal its being "made of" a number of "quanta", the electrons, whose dynamical behaviour is only "statistically" accounted for by the de Broglie-Schrödinger wave. Without entering here into the long standing debate that saw Einstein and Bohr in opposite camps, I wish to remark that it is in our "measuring" the electron beam, i.e. in our effort to find what is the beam composed of, that the discontinuous "quantum" aspects of QM reveal itself. Were we rather interested in the **collective** properties of the beam itself, thus foregoing a **short distance probe** of its composition, we would find the wave aspects perfectly adequate to describe their dynamical behaviour.

In order to understand a little better the latter point, which is focal to the approach to be illustrated in this book, we may find it useful to go back to the paradigm of classical fields, the electromagnetic (e.m.) field. Neglecting the historically important, but scientifically irrelevant, efforts that Maxwell, Kelvin and others profused to produce a mechanical model of the physical system governed by the Maxwell equations, Classical Electrodynamics (CED) stands as a great monument of a classical field system capable of describing with remarkable accuracy a multitude of wave-phenomena, involving radio waves and light beams. By analyzing the vibrational patterns in space and time of electric and magnetic fields in accordance with Maxwell equations, CED has been able to understand and predict an enormous amount of physical facts pertaining to the interaction of light with matter. The picture that CED yields of such coupled systems is clearly the classical dichotomic one: the e.m. field is a classical, continuous wave-field, while matter can be described, à la Lorentz, as large collections of point-like positive (the protons) and negative (the electrons) particles. This picture is still the one that is mostly used in radio and TV engineering, where it is found perfectly adequate. However, at the end of the XIX century this view of e.m. phenomena fell into the profound crisis of the "ultraviolet catastrophe" of black-body radiation, whose resolution required the bold hypothesis, due to Planck [1], of the discontinuity of the e.m. field, realized by the existence of elementary packets of field energy and momentum, the photons. The photoelectric effect, Compton scattering etc. further showed the deep, fundamental reality of the quanta of the e.m. field, that when probed at distances much smaller than their wave-length unmistakably exhibit a particle behaviour. If the photon is then such a real, indispensable object, how can it be made to peacefully coexist with the e.m. field of the radio engineer? How is one to attribute so different, even antithetical physical behaviours to the same physical systems? The answer to these questions is one of the great achievements of this century's theoretical physics through the discovery of the all important concept of the **Quantum Field** (QF).

Central as it is for the developments to be expounded in this book, the notion of QF fortunately can be introduced in a relatively simple way[2]. Let us consider its simplest form, that of a scalar field $\varphi(\vec{x}, t)$. According to ordinary, first-quantized QM any dynamical system must first be resolved in its fundamental degrees of freedom (the three space-positions $\vec{x}$ for a matter point) which can be "quantized" by the well known Dirac connection between Poisson parentheses and operator commutation relations. For a quantum field the procedure is essentially the same, the only (not fundamental at all) difference is that for $\varphi(\vec{x}, t)$ the resolution in the basic degrees of freedom corresponds to the expansion in a complete system of modes, that for a **local** field must contain an infinite number of degrees of freedom. The close analogy between QM and Quantum Field Theory (QFT) can be seen in the following way: for QM, the starting point is the Lagrangian

$$L = L(q_i, \dot{q}_i, t) \qquad i = 1, \ldots N, \tag{1.1}$$

where q_i is a (finite) set of Lagrangian degrees of freedom; one then defines the momenta p_i conjugate to q_i as

$$p_i = \frac{\partial L}{\partial \dot{q}_i}, \tag{1.2}$$

and the configuration space of the system is the infinite dimensional Hilbert space spanned by the operators q_i and p_i obeying the canonical commutation relations

$$[p_i, q_j] = -i\hbar \delta_{ij}, \tag{1.3}$$

where in the following the Planck constant $\hbar$ together with c, the velocity of light, and the Boltzmann constant k_B shall always be taken to be unity. For our QF $\varphi(\vec{x}, t)$

the quantization proceeds along very similar lines. One discretizes the quantization volume V as a finite lattice with lattice constant a (the limit $a \to 0$ is to be taken at the end) and identifies $\varphi(\vec{x}_j, t)$ ($\vec{x}_j$ is any of the lattice points) with the Lagrangian coordinates q_j ($j = 1, 2, \ldots, V/a^3$) . In usual classical field theories the Lagrange function is obtained from a Lagrange density $\mathcal{L}(\varphi, \dot{\varphi}, \vec{\nabla}\varphi; t)$, which is a function of φ, $\dot{\varphi}$ and its gradient $\vec{\nabla}\varphi$, and is quadratic in $\dot{\varphi}$; thus one writes ($\vec{u}$ are the lattice units vectors)

$$L = \int d^3x \mathcal{L} = \sum_{i,\vec{u}} \mathcal{L}(q_i, \dot{q}_i, q_{i+\vec{u}}; t) a^3. \tag{1.4}$$

Normalizing the field so that the quadratic term in $\dot{\varphi}$ of $\mathcal{L}$ is just $\frac{1}{2}\dot{\varphi}^2$, one obtains for the canonical momentum associated to q_i

$$p_i = \frac{\partial L}{\partial \dot{q}_i} = \dot{q}_i a^3, \tag{1.5}$$

and the canonical commutation relations read

$$[p_i, q_k] = -i\delta_{i,k}. \tag{1.6}$$

Going back to the continuum ($a \to 0$) this latter commutation relation becomes $[\Pi(\vec{x}, t) = \dot{\varphi}(\vec{x}, t)]$

$$[\Pi(\vec{x}, t), \varphi(\vec{y}, t)] = -i\delta^3(\vec{x} - \vec{y}). \tag{1.7}$$

Introducing a complete orthonormal set of modes $\{\Psi_n(\vec{x})\}$ of the field φ, solutions of some convenient eigenvalue problem (plane waves, in the case of the free e.m. field), it is a simple exercise to show that writing

$$\varphi(\vec{x}, t) = \frac{1}{\sqrt{2}} \sum_n \left[a_n(t)\Psi_n(\vec{x}) + a_n^\dagger(t)\Psi_n^*(\vec{x}) \right] \tag{1.8}$$

$$\Pi(\vec{x}, t) = \frac{-i}{\sqrt{2}} \sum_n \left[a_n(t)\Psi_n(\vec{x}) - a_n^\dagger(t)\Psi_n^*(\vec{x}) \right] \tag{1.9}$$

the equal-time commutation relation (1.7) is satisfied provided the amplitudes $a_n(t)$ obey the equal-time commutation relations

$$[a_n(t), a_m^\dagger(t)] = \delta_{nm}. \tag{1.10}$$

The mode expansion (1.8) and (1.9) and the simple "harmonic oscillator" commutation relations (1.10) for the operator field amplitudes $a_n(t)$ illustrate very well the basic nature and meaning of the physical object Quantum Field. Indeed the field amplitudes $a_n(t)$ are unitarily connected to a set of a_n's that span a huge Hilbert space consisting of the direct product of the Hilbert spaces for each "harmonic oscillator" amplitude a_n, whose non-negative quantum numbers N_n (the eigenvalues of the number operator $a_n^\dagger a_n$) can be simply understood as the "occupation numbers" of the mode $\Psi_n(\vec{x})$ or, in other words, as the number of quanta which in a given field configuration populate that particular mode.

Thus the quantum mechanical description of the field $\varphi(\vec{x}, t)$, i.e. its configuration space, can be reduced to the construction of the **Fock space**, whose base vectors are labelled by a set $\{N_n\}$ of non-negative integers, the occupation numbers of each independent mode. Accordingly any physical configuration of the quantum field $\varphi(\vec{x}, t)$ can be obtained as the linear (coherent) superposition of physical states characterized by a **well defined set of occupation numbers**, i.e. by well defined numbers of quanta, **or particles**, in well-defined states.

In this way we see very clearly that what a quantum field really describes is in general a system of a large number of identical quanta, or particles, distributed over a range of individual states, the field modes. The above discussion makes it also clear that in general the correct description of systems containing a small number of particles is **also** a second-quantized quantum field, whose dynamical evolution happens to span low multiplicity sectors of the Fock space. I believe that this observation might be of some relevance in order to clarify the nature of quantum measurements and of their alleged paradoxes, for this view would associate the first-quantized Schrödinger wave-function only to the very peculiar, and unlikely, wave-field configuration which describes an isolated (unobservable) single particle state.

Going now to the point of view of a radio-engineer: what is he to do with such a picture? How is he supposed to relate a mathematical object such as the Quantum Field, with his classical e.m. field $A_k(\vec{x}, t) = [a_k(\vec{x}, t)e^{i\Theta_k(\vec{x}, t)} + \text{c.c.}]$, described by a real amplitude $a_k(\vec{x}, t)$ and a phase $\Theta_k(\vec{x}, t)$, which evolves in space and time according to the Maxwell field equations? In order to give a proper answer to these questions, let us derive a simple and fundamental result concerning the uncertainty relations between quantum amplitudes and phases in Fock space. Let us consider the quantum amplitude a_n defined in (1.8) and (1.9), which we can write as

$$a_n = e^{i\Theta_n} N_n^{1/2}, \tag{1.11}$$

with the number operator $N_n = a_n^\dagger a_n$. Let us now insert (1.11) into the commutation relation (1.10); we have

$$[a_n, a_n^\dagger] = [e^{i\Theta_n} N_n^{1/2}, N_n^{1/2} e^{-i\Theta_n}] = e^{i\Theta_n} N_n e^{-i\Theta_n} - N_n = \mathbf{1}. \tag{1.12}$$

It is a simple exercise (Ex.$\langle 1.1 \rangle$) to show that (1.12) implies the validity of the commutation relations:

$$[\Theta_n, N_n] = -i, \tag{1.13}$$

which reveal that the phase operator Θ_n and the number operator N_n are canonically conjugate, like momentum and position in ordinary QM. It is well known from ordinary QM how from (1.13) one can derive the Heisenberg uncertainty relations

$$(\Delta\Theta_n)(\Delta N_n) \geq \frac{1}{2}, \tag{1.14}$$

whose simple consequence is that on a state with a fixed number of particles, for which $\Delta N_n = 0$, the phase is completely undetermined. Thus the states of the e.m. field that are relevant for the radio-engineer (and for the laser physicist as well) are not those consisting of a **fixed** number of photons ($\Delta N_n = 0$), for which the phase is undefined, but rather those "classical" states for which the phase is well defined and ΔN_n is accordingly undetermined. We know perfectly well what kind of states of the Fock space these correspond to: they are nothing but the "coherent states" $|\alpha_n\rangle$, eigenstates of the operator a_n, for which

$$a_n\, |\alpha_n\rangle = \alpha_n\, |\alpha_n\rangle\,. \tag{1.15}$$

As we shall see later, they are represented by infinite linear combinations of eigenstates $|n\rangle$ of the number operators N_n. The commutation relation (1.13) is just another aspect, at the higher level of quantum fields, of the quantum mechanical wave-particle duality. However, in the case of large ensembles of N identical particles ($N = \sum_n N_n$) described by a single quantum field, we may well envisage physical situations in which the relevant N_n's fluctuate around macroscopic values and for which the uncertainties ΔN_n are much smaller than N_n. It is then clear that when this occurs it is the "classical" field aspect, with its well defined phase $\Theta(\vec{x}, t)$, that becomes of definite relevance and induces us to divest our system of its quantum, "atomistic" features, to reach a global, "collective" description of its dynamical evolution in terms of a "classical" (complex) wave-field $\varphi(\vec{x}, t)$. In this way $\varphi(\vec{x}, t)$ becomes a kind of **macroscopic** Schrödinger wave-function that accounts for the genuinely collective dynamical behaviour of the large ensemble of identical particles. Enmeshed in a huge web of phase relationships, the individual mass points are thus seen to completely lose their individuality, to become involved in a well-defined **coherent** dynamical behaviour, completely specified by the classical wave-field $\varphi(\vec{x}, t)$. And if we go through the intellectual path that led from the classical Maxwell field to Planck's photons

backwards, we will perceive the physical meaning of the collective, macroscopic classical wave-field, much in the same way as the radio-engineer understands his classical
e.m. wave-field, with well defined amplitudes, intensities and phases, without worry-
ing about the photon content of its oscillating electric and magnetic fields: for him
the notion of photon is just too remote from the physical reality he is interested in
to be of any **real** importance.

Before analyzing the quantum structure and kinematics of the most relevant wave-
fields of Condensed Matter (CM), a comment on the mode expansion (1.8) which led
to the introduction of the Fock space. Actually one should speak of a Fock space, for
its construction utilizes the creation (annihilation) operators $a_n^\dagger$ (a_n), that appear as
the coefficients of the mode expansion (1.8). Should we consider expanding $\varphi(\vec{x}, t)$
over a different orthonormal system $\{\chi_m(\vec{x})\}$ the associated coefficients $b_m^\dagger$ (b_m) would
also obey commutation relations of the type (1.10) and the new Fock space spanned
by them would correspond to a well defined unitary "rotation" of the original one.
Thus an important practical question is what expansion and what Fock space are best
suited to describe the space of quantum configurations of the quantum field $\varphi(\vec{x}, t)$.
It should be clear that the answer to this question must take the actual structure of
the Hamiltonian operator into consideration. As a general rule the best expansion is
the one for which the Hamiltonian is diagonal or only slightly (perturbatively) non-
diagonal in its associated Fock space. Thus the **art** of introducing a mode expansion
for a QF consists in identifying an orthonormal system whose Fock space makes the
Hamiltonian as closely diagonal as possible. For free-fields the solution is completely
straightforward; in the general case, as we shall see, a solution can only be obtained
when a sufficient understanding of the Ground State (GS) has been achieved.

1.2   The Electromagnetic Quantum Field

We shall now apply the general considerations of the preceding Section to the
"key actor of the drama of Condensed Matter": the e.m. quantum field [3].

The starting point is, as always, classical electromagnetism which is compactly
described by the Lagrangian density $[x = (x_o, \vec{x})]$

$$\mathcal{L}(x) = -\frac{1}{4}F_{\mu\nu}(x)F^{\mu\nu}(x), \tag{1.16}$$

where the field strength tensor can be obtained from the vector potential $A_\mu(x)$

$$F_{\mu\nu}(x) = \partial_\mu A_\nu(x) - \partial_\nu A_\mu(x). \tag{1.17}$$

Recall that the electric $\vec{E}$ and magnetic $\vec{B}$ fields are related to the components of $F_{\mu\nu}$

as

$$E_i = F_{oi}, \tag{1.18}$$

and

$$B_i = \frac{1}{2}\epsilon_{ijk}F_{jk}. \tag{1.19}$$

The principle of extremal action as applied to the action of the e.m. field (i.e. the space-time integral of the Lagrangian density (1.16)) produces the free e.m. field equations:

$$\partial^\mu F_{\mu\nu}(x) = 0, \tag{1.20}$$

or in terms of the vector potentials

$$\partial^\mu \partial_\mu A_\nu - \partial_\nu \partial^\mu A_\mu = 0, \tag{1.21}$$

which reduce to the d'Alembert equation "in the gauge" $\partial^\mu A_\mu = 0$. As is well known the possibility of "choosing a gauge", i.e. imposing to A_μ a particular constraint $F(A) = 0$, is afforded by the invariance of the Maxwell Lagrangian (1.16) under the inhomogeneous gauge-transformations

$$A_\mu(x) \to A_\mu(x) + \partial_\mu \lambda(x), \tag{1.22}$$

$\lambda(x)$ being any properly well behaved space-time function. An equally well known consequence of the invariance under the gauge transformations (1.22) is that the vector potentials A_μ provide a redundant description of the e.m. field. As a result of imposing a gauge, such as the Lorentz gauge $\partial^\mu A_\mu = 0$, the number of degrees of freedom per space-point gets reduced from 4 to 3. But for the free e.m. field, described by the Lagrangian density (1.16), choosing a given space-time gauge does not completely exhaust the redundancy of $A_\mu(x)$. Indeed, once for instance the "time-like" gauge $A_o(x)$ has been chosen, using (1.22) we can obtain a vector potential that is "transverse", i.e. satisfies

$$\vec{\nabla} \cdot \vec{A} = 0. \tag{1.23}$$

through a further time-independent gauge-function $\lambda(\vec{x})$.

    Deprived of its "longitudinal" component $\vec{\nabla} \cdot \vec{A}$, the vector potential $\vec{A}$ finally describes a system that possesses only two degrees of freedom per space-point: the two polarizations of the transverse e.m. field.

Before considering the quantization of $A_\mu(\vec{x}, t)$ in the gauge $A_o = 0$ and $\vec{\nabla} \cdot \vec{A} = 0$, the so-called "radiation gauge", let us note two important conserved operators that can be derived from the Maxwell Lagrangian by use of the Nöther theorem: the Hamiltonian (energy) operator:

$$H = \frac{1}{2} \int (\vec{E}^2 + \vec{B}^2) d^3x, \qquad (1.24)$$

and the momentum operator

$$P_k = \frac{1}{2} \epsilon_{klm} \int E_l B_m d^3x. \qquad (1.25)$$

Taking now $A_j(\vec{x}, t)$ as our (not all independent) quantum fields, and recalling that the Maxwell Lagrangian is

$$L = \int d^3x \, \mathcal{L} = \frac{1}{2} \int d^3x \, (\dot{A}_j \dot{A}_j - \partial_j A_k \partial_j A_k) \qquad (1.26)$$

we immediately see that the momentum conjugate to A_j is just $\dot{A}_j = E_j$, so that we would be tempted to write the canonical equal-time commutation relations, analogous to (1.7), as

$$[E_j(\vec{x}, t), A_k(\vec{y}, t)] = -i \delta_{jk} \delta^3(\vec{x} - \vec{y}). \qquad (1.27)$$

These commutation relations would however be inconsistent with the transversality condition $\vec{\nabla} \cdot \vec{A} = 0$. The way around this difficulty is well known: one introduces as the relevant orthonormal system that of the transverse solutions of the d'Alembert equation in a box of volume $V = L^3$, with periodic boundary conditions ($\vec{k} = (2\pi/L)\vec{n}$, n_i integers; $r = 1, 2$ specifies the two independent polarizations)

$$\vec{f}_{\vec{k}r}(\vec{x}, t) = \frac{1}{\sqrt{2\omega_{\vec{k}} V}} e^{-i(\omega_{\vec{k}} t - \vec{k} \cdot \vec{x})} \vec{\epsilon}_{\vec{k}r}, \qquad (1.28)$$

with

$$\omega_{\vec{k}} = |\vec{k}|, \qquad (1.29)$$

$$\vec{k} \cdot \vec{\epsilon}_{\vec{k}r} = 0, \qquad (1.30)$$

and

$$\vec{\epsilon}_{\vec{k}r} \cdot \vec{\epsilon}_{\vec{k}s} = \delta_{rs}. \qquad (1.31)$$

One then sets

$$A_j(\vec{x},t) = \sum_{\vec{k}r}(a_{\vec{k}r}f_{j,\vec{k}r}(\vec{x},t) + a_{\vec{k}r}^\dagger f_{j,\vec{k}r}(\vec{x},t)^*) \qquad (1.32a)$$

$$E_j(\vec{x},t) = -i\omega_{\vec{k}}\sum_{\vec{k}r}(a_{\vec{k}r}f_{j,\vec{k}r}(\vec{x},t) - a_{\vec{k}r}^\dagger f_{j,\vec{k}r}(\vec{x},t)^*) \qquad (1.32b)$$

where $a_{\vec{k}r}$ $(a_{\vec{k}r}^\dagger)$ are operator Fourier coefficients. One can now compute the Hamiltonian operator: From (1.24) and the orthogonality properties of the plane waves one easily obtains (Ex. $\langle 1.2 \rangle$):

$$H = \frac{1}{2}\sum_{\vec{k}r}\omega_{\vec{k}}(a_{\vec{k}r}^\dagger a_{\vec{k}r} + a_{\vec{k}r}a_{\vec{k}r}^\dagger), \qquad (1.33)$$

which shows that for each mode $\vec{f}_{\vec{k}r}(\vec{x},t)$, the dynamics is that of an harmonic oscillator with amplitude

$$Q_{\vec{k}r} = \frac{1}{\sqrt{2}}(a_{\vec{k}r} + a_{\vec{k}r}^\dagger), \qquad (1.34)$$

and conjugate momentum

$$P_{\vec{k}r} = \frac{i}{\sqrt{2}}(a_{\vec{k}r}^\dagger - a_{\vec{k}r}). \qquad (1.35)$$

Thus the canonical commutation relations hold for $Q_{\vec{k}r}$ and $P_{\vec{k}r}$ provided

$$[a_{\vec{k}r}(t), a_{\vec{k}'s}^\dagger(t)] = \delta_{rs}\delta_{\vec{k}\vec{k}'}, \qquad (1.36)$$

which by use of (1.32) lead to the equal-time commutation relations:

$$[E_j(\vec{x},t), A_k(\vec{y},t)] = -i\delta_{jk}^{tr}(\vec{x} - \vec{y}), \qquad (1.37)$$

where $\delta_{jk}^{tr}(\vec{x}) = \int \frac{d^3k}{(2\pi)^3}[\delta_{jk} - (k_j k_k/\vec{k}^2)]e^{i\vec{k}\cdot\vec{x}}$ is the non local "transverse delta function". All this may seem rather contrived, however the correctness of (1.37) is really based upon the equal-time commutation relations (1.36) which properly define the Fock space for **each independent degree of freedom** of the free e.m. field. Using these commutation relations the Hamiltonian operator may be rewritten as

$$H = \sum_{\vec{k}r}\omega_{\vec{k}}(a_{\vec{k}r}^\dagger a_{\vec{k}r} + \frac{1}{2}), \qquad (1.38)$$

and analogously, for the momentum operator (1.25) one obtains

$$\vec{P} = \sum_{\vec{k}r} \vec{k}(a^\dagger_{\vec{k}r} a_{\vec{k}r} + \frac{1}{2}). \tag{1.39}$$

We thus see that the configuration space of the quantized free e.m. radiation (transverse) field simply corresponds to the Hilbert space associated to the "ether oscillators", with amplitudes $a_{\vec{k}r}$, one for each dynamically independent degree of freedom of the classical radiation field. The simple form of the Hamiltonian (1.38) shows that the ground state is the state where the occupation numbers, eigenvalues of the operators $N_{\vec{k}r} = a^\dagger_{\vec{k}r} a_{\vec{k}r}$, are all zero, and its energy is

$$E_{vac} = \sum_{\vec{k}r} \frac{\omega_{\vec{k}}}{2}, \tag{1.40}$$

i.e. the sum of the zero point energies of each "ether oscillator", demonstrating that the free quantized e.m. field in its ground state possesses an enormous energy. Can this energy be tapped? We shall see that a positive answer to this question will always imply strong interactions of the e.m. field with charged matter which will cause the free ground state (which we may also call the "perturbative" ground state (PGS), or vacuum) to be unstable.

To end this Section, let us give some attention to the problem of the radio-engineer. How is he going to relate his measured radiation field $A_j(\vec{x}, t)_c$ in free space to the quantum field that we have discussed? In other words what is the form of the Fock space vector $\left| \vec{A}(\vec{x}, t) \right\rangle$ for which

$$\left\langle \vec{A}(\vec{x}, t)_c \right| A_j(\vec{x}, t) \left| \vec{A}(\vec{x}, t)_c \right\rangle = A_j(\vec{x}, t)_c \ ? \tag{1.41}$$

The answer is really quite easy. Let us expand the classical field in the plane wave basis (1.28)

$$A_j(\vec{x}, t)_c = \sum_{\vec{k}r} [\alpha_{\vec{k}r}(t) f_{j,\vec{k}r}(\vec{x}, t) + \alpha_{\vec{k}r}(t)^* f_{j,\vec{k}r}(\vec{x}, t)^*], \tag{1.42}$$

then the state $\left| \vec{A}(\vec{x}, t)_c \right\rangle$ is simply given by

$$\left| \vec{A}(\vec{x}, t)_c \right\rangle = \prod_{\vec{k}r} \left| \alpha_{\vec{k}r}(t) \right\rangle_{\vec{k}r} \tag{1.43}$$

where the "coherent state" $\left| \alpha(t) \right\rangle_{\vec{k}r}$ is a normalized eigenvector of the annihilation operator $a_{\vec{k}r}(t)$, with eigenvalue $\alpha_{\vec{k}r}(t)$, for which one has

$$a_{\vec{k}r} \left| \alpha(t) \right\rangle_{\vec{k}r} = \alpha_{\vec{k}r}(t) \left| \alpha(t) \right\rangle_{\vec{k}r}. \tag{1.44}$$

The proof of this is simple, for taking account of (1.32a) we have

$$
\left\langle \vec{A}(\vec{x},t)_c \middle| A_j(\vec{x},t) \middle| \vec{A}(\vec{x},t)_c \right\rangle
$$
$$
= \left\langle \vec{A}(\vec{x},t)_c \middle| \sum_{\vec{k}r}[a_{\vec{k}r}(t)f_{j,\vec{k}r}(\vec{x},t) + a_{\vec{k}r}^\dagger(t)f_{j,\vec{k}r}(\vec{x},t)^*] \middle| \vec{A}(\vec{x},t)_c \right\rangle, \tag{1.45}
$$

and from the straightforward consequences of (1.44)

$$
_{\vec{k}r}\left\langle \alpha_{\vec{k}r}(t) \middle| a_{\vec{k}r} \middle| \alpha_{\vec{k}r}(t) \right\rangle_{\vec{k}r} = \alpha_{\vec{k}r}(t) \tag{1.46a}
$$

and

$$
_{\vec{k}r}\left\langle \alpha_{\vec{k}r}(t) \middle| a_{\vec{k}r}^\dagger \middle| \alpha_{\vec{k}r}(t) \right\rangle_{\vec{k}r} = \alpha_{\vec{k}r}(t)^*, \tag{1.46b}
$$

we can rewrite (1.45)

$$
\left\langle \vec{A}(\vec{x},t) \middle| A_j(\vec{x},t) \middle| \vec{A}(\vec{x},t) \right\rangle = \sum_{\vec{k}r}[\alpha_{\vec{k}r}(t)f_{j,\vec{k}r}(\vec{x},t) + \alpha_{\vec{k}r}^*(t)f_{j,\vec{k}r}(\vec{x},t)^*], \tag{1.47}
$$

which coincides with (1.41).

But to give our radio-engineer full satisfaction we must also disclose the relationship between $\left|\alpha_{\vec{k}r}(t)\right\rangle_{\vec{k}r}$ and the Fock space base vectors $\left|n_{\vec{k}r}\right\rangle_{\vec{k}r}$ that record the simultaneous presence of $n_{\vec{k}r}$ photons in the plane wave mode with momentum $\vec{k}$ and transverse polarization $\vec{\epsilon}_{\vec{k}r}$. Dropping for convenience the indices $\vec{k}$ and r, we write

$$
|\alpha\rangle = \sum_{n=0}^{\infty} c_n |n\rangle, \tag{1.48}
$$

the eigenvalue equation (1.44) implies

$$
a |\alpha\rangle = \sum_{n=1}^{\infty} c_n \sqrt{n} |n-1\rangle = \alpha \sum_{n=1}^{\infty} c_n |n\rangle. \tag{1.49}
$$

Using the orthonormality of the harmonic oscillator states $|n\rangle$ we obtain the simple recursion relation:

$$
c_{n+1} = \frac{\alpha(t)}{\sqrt{n+1}} c_n, \tag{1.50}
$$

whose solution is

$$
c_n = \frac{\alpha(t)^n}{\sqrt{n!}} c_o. \tag{1.51}
$$

Using the normalization condition we can finally write

$$|\alpha\rangle = \exp(-\frac{\alpha^*\alpha}{2}) \sum_{n=0}^{\infty} \frac{\alpha^n}{\sqrt{n!}} |n\rangle, \qquad (1.52)$$

which, in accordance with our expectations from the indeterminacy relation (1.14), shows that the eigenvector of the "annihilation" operator a must involve a linear combination of an infinite number of eigenvectors of the number operator $N = a^\dagger a$. Let us now compute the dispersion ΔN of the number operator in such coherent states: one has

$$(\Delta N)^2 = \langle\alpha| (a^\dagger a)^2 |\alpha\rangle - \langle\alpha| a^\dagger a |\alpha\rangle^2 = \langle\alpha| a^\dagger a |\alpha\rangle = |\alpha|^2 = \langle N\rangle, \qquad (1.53)$$

i.e. the "photons" are Poisson distributed around their mean number $\langle N\rangle = |\alpha|^2$. An interesting feature of the coherent states $|\alpha\rangle$ is that they are not orthogonal. Indeed one has

$$\langle\alpha'|\alpha\rangle = \sum_{n,n'} \langle n'| e^{-|\alpha'|^2/2} e^{-|\alpha|^2/2} |n\rangle \frac{\alpha'^{*n'}\alpha^n}{\sqrt{n!n'!}}$$
$$= \exp\left[-\frac{1}{2}|\alpha' - \alpha|^2\right] \exp\left[\frac{1}{2}(\alpha'\alpha^* - \alpha'^*\alpha)\right], \qquad (1.54)$$

i.e. they form what is called an "overcomplete system". Finally we record an useful formula for the resolution of the identity, which reads

$$\frac{-i}{\pi} \int d^2\alpha \, |\alpha\rangle\langle\alpha| = \sum_n |n\rangle\langle n|, \qquad (1.55)$$

where the integration is over the complex plane ($\alpha = \alpha_1 + i\alpha_2$, $d^2\alpha = id\alpha_1 d\alpha_2$) (Ex. $\langle 1.3\rangle$).

1.3   The matter wave-field

In this Section we shall follow explicitly the path that from a collection of N identical quantum mechanical systems contained in a volume V leads to the kinematics and the dynamics of the matter quantum field, or simply wave-field, associated with such collection[4]. In Condensed Matter physics it is customary to write the Hamiltonian H_{matt} for such system in the many particle formalism

$$H_{matt} = \sum_{j=1}^{N} H_o(\vec{x}_j, \vec{p}_j, \alpha_j) + V(\vec{x}_1, \alpha_1, \ldots, \vec{x}_N, \alpha_N), \qquad (1.56)$$

where $H_o(\vec{x}, \vec{p}, \alpha)$ is the single-particle Hamiltonian, function of the center of mass coordinate of the mechanical system $\vec{x}$, its total momentum $\vec{p}$, and of the relevant internal coordinates α, necessary to describe the internal structure of the elementary system. H_o includes all interactions with external fields as well as with "self consistent" Hartree-Fock type potentials, that preserve the single particle character of the dynamical evolution . The potential term V customarily describes the residual short-range interactions among the systems and is usually written as the sum of two or few body interaction terms. Leaving the short-range interaction aside for the moment, let us analyse the "free" matter system described by the first term of (1.56). Due to its factorized dynamics it is clear that instead of following the dynamics of each one of the elementary systems, we may as well describe its dynamical evolution by counting how many of them are to be found in each of the stationary states of the Hamiltonian H_o, i.e. by specifying, as in the case of the photons of the e.m. field, the "occupation numbers" of the different eigenstates of the Hamiltonian. Let $\{\varphi_n(\vec{x}, \alpha)\}$ be the complete orthonormal system that diagonalizes H_o, and let the relative eigenvalues be denoted by ϵ_n such that

$$H_o\varphi_n(\vec{x}, \alpha) = \epsilon_n\varphi_n(\vec{x}, \alpha). \qquad (1.57)$$

We can now associate to each "single particle" level $\{\varphi_n(\vec{x}, \alpha)\}$ an annihilation operator a_n, such that $N_n = a_n^\dagger a_n$ specifies the occupation number of the level $\{\varphi_n(\vec{x}, \alpha)\}$ obeying the (anti)commutation relations

$$[a_n, a_m^\dagger]_\pm = \delta_{nm}, \qquad (1.58)$$

according to whether our identical systems have Fermi or Bose character. It it clear that in the Fock space spanned by $\{a_n, a_n^\dagger\}$ the free Hamiltonian H_o can be expressed as

$$H_o = \sum_{j=1}^{N} H_o(\vec{x}_j, \vec{p}_j, \alpha_j) = \sum_n \epsilon_n a_n^\dagger a_n, \qquad (1.59)$$

whose eigenvalues are given by

$$E_{\{n_m\}} = \sum_m n_m \epsilon_m. \qquad (1.60)$$

We now write down the quantum field:

$$\Psi(\vec{x}, \alpha; t) = \sum_n a_n(t)\varphi_n(\vec{x}, \alpha), \qquad (1.61)$$

where $a_n(0) = a_n$, and the operator set $\{a_n(t)\}$ is unitarily related to $\{a_n\}$ and the equal-time (anti)commutation relations

$$[a_n(t), a_m^\dagger(t)]_\pm = \delta_{nm}, \tag{1.62}$$

are thus valid at any time t. Using the orthonormality of the system $\{\varphi_n(\vec{x}, \alpha)\}$ we can immediately write

$$H_o = \sum_n \epsilon_n a_n^\dagger a_n = \int_{\vec{x},\alpha} \Psi^\dagger(\vec{x}, \alpha; t) H_o(\vec{x}, \alpha) \Psi(\vec{x}, \alpha; t), \tag{1.63}$$

having thus converted the discrete sum over N single particle Hamiltonians into a field theoretical expression involving the wave-field $\Psi(\vec{x}, \alpha; t)$ and its hermitian conjugate: an operation which apparently has led to the disappearance of the particle number N. Before searching for it, let us compute the equal-time (anti)commutation relations of the wave-field with its hermitian conjugate. One has

$$[\Psi(\vec{x}, \alpha; t), \Psi^\dagger(\vec{y}, \beta; t)]_\pm = \sum_{nm} \varphi_n(\vec{x}, \alpha) \varphi_m^*(\vec{y}, \beta) [a_n(t), a_m^\dagger(t)]_\pm = \delta(\vec{x} - \vec{y}) \delta_{\alpha\beta}, \tag{1.64}$$

which follows from the (anti)commutation relations (1.62) and the completeness of the orthonormal system $\{\varphi_n(\vec{x}, \alpha)\}$. Eq. (1.64) is very important for it shows that the field $i\Psi^\dagger(\vec{x}, \alpha)$ is the canonical conjugate momentum operator to $\Psi(\vec{x}, \alpha)$. This allows us to write the Lagrangian of the "free" matter system as the simple "Legendre-transform" of the Hamiltonian H_o, i.e.

$$L_{matt} = i \int_{\vec{x},\alpha} \Psi^\dagger(\vec{x}, \alpha; t) \frac{\partial}{\partial t} \Psi(\vec{x}, \alpha; t) - H_o, \tag{1.65}$$

which is obviously invariant under the phase transformations $\Psi(\vec{x}, \alpha) \to e^{ix}\Psi(\vec{x}, \alpha)$, $\Psi^\dagger(\vec{x}, \alpha) \to e^{-ix}\Psi^\dagger(\vec{x}, \alpha)$. With the aid of the Nöther theorem this implies that the operator

$$\hat{N} = \int_{\vec{x},\alpha} \Psi^\dagger(\vec{x}, \alpha; t) \Psi(\vec{x}, \alpha; t) \tag{1.66}$$

is conserved. Using again the orthonormality of the system $\{\varphi_n(\vec{x}, \alpha)\}$ we may write

$$\hat{N} = \sum_n a_n^\dagger a_n, \tag{1.67}$$

revealing that the eigenvalues of $\hat{N}$ coincide with the number of elementary systems contained in all the different modes. Thus our search has not been very long: the number of particles N existing in our volume V just labels the subspace of the Fock space in which, due to the conservation of $\hat{N}$, the dynamical evolution of the wave-field takes place.

Going back to (1.66) it does not take too much to realize that the operator

$$\rho(\vec{x},t) = \int_\alpha \Psi^\dagger(\vec{x},\alpha;t)\Psi(\vec{x},\alpha;t), \tag{1.68}$$

is nothing but the particle density operator, which in the many particle formalism has the expression

$$\rho(\vec{x},t) = \sum_{j=1}^N \delta^3(\vec{x} - \vec{x}_j(t)) \tag{1.69}$$

Use of (1.68) and (1.69) allows us to give a quantum field theoretical expression to any short-range interaction term V. For instance, in the case of a typical two-body interaction potential one may write:

$$V^{(2)} = \sum_{i<j}^N V(\vec{x}_i - \vec{x}_j) = \frac{1}{2}\int dx \int dy \sum_{i,j}^N \delta^3(\vec{x} - \vec{x}_i)\delta^3(\vec{y} - \vec{x}_j)V(\vec{x} - \vec{y})$$

$$= \frac{1}{2}\int dx \int dy \rho(\vec{x},t)V(\vec{x} - \vec{y})\rho(\vec{y},t) \tag{1.70}$$

$$= \frac{1}{2}\int_{\vec{x},\alpha}\int_{\vec{y},\beta} \Psi^\dagger(\vec{x},\alpha;t)\Psi(\vec{x},\alpha;t)V(\vec{x} - \vec{y})\Psi^\dagger(\vec{y},\beta;t)\Psi(\vec{y},\beta;t);$$

and the generalization to many-body interactions and to potentials that have a non-trivial dependence on the internal variables α is completely straightforward.

As explained in the Foreword the basic lacuna that this book wishes to fill in the analysis of Condensed Matter systems is the generalized neglect of the electro-dynamical interaction of the elementary systems with the radiation field $\vec{A}(\vec{x},t)$, we end thus this Section with a preliminary analysis of the type of $\vec{A}$-dependent terms that may occur in the matter Hamiltonian H_{matt}. From (1.65) it is quite clear that $\vec{A}$-dependent terms will arise from the minimal interaction obtained by "minimally" shifting the momenta of the charges as

$$\vec{p}_i \to \vec{p}_i + e_i\vec{A}(\vec{x}_i,t). \tag{1.71}$$

The dependence of the Hamiltonian H_o being in general quadratic in such momenta, the shift (1.71) will generate terms linear and quadratic in $\vec{A}$, that in view of the very small (with respect to the relevant e.m. wave-lengths) size of the elementary systems can be written in the general forms (e is the normalized electron charge)

$$H_{rad}^{(1)} = e \int_{\vec{x},\alpha} \vec{A}(\vec{x},t) \cdot \Psi^\dagger(\vec{x},\alpha;t)\vec{J}(\alpha)\Psi(\vec{x},\alpha;t) \tag{1.72}$$

and

$$H_{rad}^{(2)} = e^2\lambda \int_{\vec{x},\alpha} \vec{A}(\vec{x},t)^2 \Psi^\dagger(\vec{x},\alpha;t)\Psi(\vec{x},\alpha;t) \tag{1.73}$$

where the operator $\vec{J}(\alpha)$ and the constant λ depend on the particular structure of the elementary systems under consideration. It is appropriate to remark at this point that the constant λ is not given simply by the substitution of the minimal shift (1.71) in the "free" Hamiltonian H_o, for we must include in the interaction Hamiltonian $H_{rad}^{(2)}$ all terms that are second order in $\vec{A}$. It requires very little thought to realize that terms of this type are also generated by the second order iteration of $H_{rad}^{(1)}$ that is well known to give rise to important dispersive forces, like the Van der Waals forces, whose physical relevance cannot be disputed. We shall see the importance of this observation in the discussion of the existence of the Hepp-Lieb superradiant phase transition in Chapter 3.

We are now in a position to write down the full structure of the Hamiltonian operator H_{matt} of the matter wave-field:

$$H_{matt} = \int_{\vec{x},\alpha} \Psi^\dagger(\vec{x},\alpha;t)H_o(\vec{x},-i\vec{\nabla},\alpha)\Psi(\vec{x},\alpha;t) + H_{rad}^{(1)} + H_{rad}^{(2)} + H_{SR}, \tag{1.74}$$

where $H_{rad}^{(1)}$ and $H_{rad}^{(2)}$ have the forms (1.72) and (1.73) respectively and H_{SR}, the short-range, generally electrostatic interaction Hamiltonian, comprises terms of the type (1.70). The Lagrangian

$$L_{matt} = i \int_{\vec{x},\alpha} \Psi^\dagger(\vec{x},\alpha;t)\frac{\partial}{\partial t}\Psi(\vec{x},\alpha;t) - H_{matt} \tag{1.75}$$

yields then the complete QFT of the matter wave-field that we have been seeking.

1.4 Interacting fields and "coherence"

Postponing to the next Chapters the analysis of the dynamics stemming from the interaction between the quantized radiation field $\vec{A}(\vec{x},t)$ and the matter field $\Psi(\vec{x},\alpha;t)$, we end this introductory "kinematical" Chapter with a brief discussion of the physic potentialities of the quantum field theoretical framework discussed above. The concept that will play a major role throughout this book is that of **coherence**, the fundamental property of a quantum field whose individual elementary objects (its quanta) participate in a collective, coherent dynamical evolution. Coherence is thus seen to be intimately related to the appearance of long range order which is signalled, for instance, by the correlation function $\langle \Omega | \ \Psi^{\dagger}(\vec{x}, \alpha; 0) \Psi(\vec{y}, \beta; 0) \ | \Omega \rangle$ ($| \Omega \rangle$ is the state of the quantum field) being non zero at large distances $|\vec{x} - \vec{y}|$. However, one can see very simply that coherence, that we may characterize by the existence of a non-zero "order parameter" (or wave function)

$$\varphi_o(\vec{x}, \alpha; t) = \langle \Omega | \ \Psi(\vec{x}, \alpha; t) \ | \Omega \rangle , \qquad (1.76)$$

is not identical with long range order. Indeed, a coherent system obeying (1.76) is certainly long-range for, if a non-zero order parameter exists, one has

$$\langle \Omega | \ \Psi^{\dagger}(\vec{x}, \alpha; 0) \Psi(\vec{y}, \beta; 0) \ | \Omega \rangle = \varphi_o^*(\vec{x}, \alpha) \varphi_o(\vec{y}, \beta), \qquad (1.77)$$

which clearly does not vanish for $|\vec{x} - \vec{y}|$ large. However a system can be long-range ordered **without** being coherent. A paradigmatic example of this fact is the Bose gas, i.e. a collection of N elementary Bose systems, whose field Hamiltonian is simply given by the "free" term H_o. The third principle of Thermodynamics, Nernst's heat theorem, demands that at zero temperature the Bose gas be in a single quantum state, its ground state. Denoting by $n = 0$ the "single particle state" of lowest energy ϵ_o and $\varphi_o(\vec{x}, \alpha)$ its wave function, the ground state $| \Omega \rangle$ of the Bose gas is clearly

$$| \Omega \rangle = | N \rangle_o \prod_{n \neq 0} | 0 \rangle_n . \qquad (1.78)$$

Such a Bose condensate is evidently long-range ordered, for the correlation function is

$$\langle \Omega | \ \Psi^{\dagger}(\vec{x}, \alpha; 0) \Psi^{\dagger}(\vec{y}, \beta; 0) \ | \Omega \rangle = \varphi_o^*(\vec{x}, \alpha) \varphi_o(\vec{y}, \beta). \qquad (1.79)$$

However this Bose condensate is not coherent, for the "order parameter"

$$\langle \Omega | \ \Psi(\vec{x}, \alpha; 0) \ | \Omega \rangle = \frac{\varphi_o(\vec{x}, \alpha)}{\sqrt{N}} \ _o\langle N | \ a_o \ | N \rangle_o , \qquad (1.80)$$

vanishes on account of the "sharply" fixed number N of non-interacting bosons, and in agreement with the uncertainty principle (1.14). This simple result shows in rather general fashion that an ordered system (which in the case of the Bose condensate can be said to be "ordered by default") may well show no trace of the "phase" coherence (1.76) that characterizes coherent systems such as superfluids and superconductors. In particular the basically non-interacting picture that underlies the notion of Bose condensation is seen to be inadequate to yield the physical properties that belong to coherent quantum field systems. Ground states of the type (1.78) represent the states of minimum energy of QFT's whose Hamiltonians are "gaussian", i.e. linear in the wave-field Ψ and its conjugate momentum $i\Psi^\dagger$. Their long-range order without phase coherence, characteristic of their free nature, will be preserved in the perturbative ground state (PGS) of a QFT that is obtained from the "gaussian theory" by a small perturbing Hamiltonian H_{int}. The reason for this is that a PGS consists in adding to (1.78) a series of Fock space vectors whose coefficients are proportional to increasing powers of the perturbative coupling constant. On the other hand, as we have seen in Section (1.3), in order to satisfy (1.76) one needs infinite superpositions of Fock space vectors, confirming that coherence is a basically non-perturbative notion.

As an illustration of the above discussion let us consider the simple case of a system of N pointlike atoms of mass m in a volume V with a short-range δ-like two-body interaction. The Hamiltonian is:

$$H = \int_{\vec{x}} \Psi^\dagger(\vec{x}, t)(-\frac{\vec{\nabla}^2}{2m})\Psi(\vec{x}, t) + \frac{\lambda}{2}\int_{\vec{x}} \Psi^\dagger(\vec{x}, t)\Psi(\vec{x}, t)\Psi^\dagger(\vec{x}, t)\Psi(\vec{x}, t), \qquad (1.81)$$

and we expand the wave-field $\Psi(\vec{x}, t)$ in plane waves in the box of volume V with periodic boundary conditions:

$$\Psi(\vec{x}, t) = \frac{1}{\sqrt{V}}\sum_{\vec{k}} a_{\vec{k}}(t)e^{i\vec{k}\cdot\vec{x}}. \qquad (1.82)$$

Our aim now is to find out whether a state of the type

$$|\Omega\rangle = |\varphi\rangle_o \, |\Omega'\rangle \qquad (1.83)$$

(where $|\varphi\rangle_o$ is a coherent state with (complex) amplitude φ, and $|\Omega'\rangle$ is a vector of the subspace of the Fock space with $\vec{k} \neq 0$) diagonalizes H, in what circumstances it is the state of minimum energy and what is the spectrum of excitations. Evidently our problem reduces to diagonalizing H on the subspace $\vec{k} \neq 0$ of the Fock space with the wave-field (we set $t = 0$)

$$\Psi = \varphi + \frac{1}{\sqrt{V}}\sum_{\vec{k}\neq 0} a_{\vec{k}}e^{i\vec{k}\cdot\vec{x}}. \qquad (1.84)$$

A simple exercise converts (1.81) into

$$H = \sum_{\vec{k} \neq 0} \frac{\vec{k}^2}{2m} a_{\vec{k}}^\dagger a_{\vec{k}} + \frac{\lambda}{2} V |\varphi|^4$$

$$+ \lambda \sum_{\vec{k}} (a_{\vec{k}}^\dagger a_{\vec{k}} |\varphi|^2 + \frac{\varphi^2}{2} a_{\vec{k}}^\dagger a_{-\vec{k}}^\dagger + \frac{\varphi^{*2}}{2} a_{\vec{k}} a_{-\vec{k}}) + H^{(3)} + H^{(4)},$$

(1.85)

where $H^{(3)}$ and $H^{(4)}$ contain three and four $a_{\vec{k}}$ $(a_{\vec{k}}^\dagger)$ operators and will be treated perturbatively, which means that at this stage they will be neglected. The diagonalization of the quadratic part of (1.85) was achieved a long time ago by the late N.N. Bogoliubov[5], with a typical feat of ingenuity that is worth following in detail. One seeks a "Bogoliubov transformation"

$$A_{\vec{k}} = \beta_{\vec{k}} a_{\vec{k}} + \gamma_{\vec{k}} a_{-\vec{k}}^\dagger$$
$$A_{\vec{k}}^\dagger = \beta_{\vec{k}}^* a_{\vec{k}}^\dagger + \gamma_{\vec{k}}^* a_{-\vec{k}}$$

(1.86)

defining new annihilation (creation) operators $A_{\vec{k}}$ $(A_{\vec{k}}^\dagger)$ that bring our quadratic Hamiltonian into the form:

$$H^{(2)} = \sum_{\vec{k} \neq 0} [a_{\vec{k}}^\dagger a_{\vec{k}} (\frac{\vec{k}^2}{2m} + \lambda |\varphi|^2) + \frac{\lambda}{2} (\varphi^{*2} a_{\vec{k}} a_{-\vec{k}} + \varphi^2 a_{\vec{k}}^\dagger a_{\vec{k}}^\dagger)]$$

$$= \sum_{\vec{k}} E_{\vec{k}} A_{\vec{k}}^\dagger A_{\vec{k}} + \text{constant}.$$

(1.87)

First one notes that the standard commutation relations allow one to set $\beta_{\vec{k}} = \cosh \alpha_{\vec{k}} e^{i\chi_{\vec{k}}}$ and $\gamma_{\vec{k}} = \sinh \alpha_{\vec{k}} e^{i\chi_{\vec{k}}}$; then by substituting (1.86) in (1.87) and comparing one easily obtains:

$$E_{\vec{k}} \cosh 2\alpha_{\vec{k}} = \frac{\vec{k}^2}{2m} + \lambda |\varphi|^2,$$

$$E_{\vec{k}} \sinh 2\alpha_{\vec{k}} = \lambda |\varphi|^2,$$

(1.88)

the phase $\chi_{\vec{k}}$ being equal to the phase of the amplitude φ. From (1.88) one obtains the energy spectrum

$$E_{\vec{k}} = \sqrt{\frac{(\vec{k}^2)^2}{4m^2} + 2\lambda |\varphi|^2 \frac{\vec{k}^2}{2m}}.$$

(1.89)

Computing the energy of the state $|\Omega\rangle$ with $|\Omega'\rangle$ annihilated by $A_{\vec{k}}$, we find

$$E(\Omega) = \frac{\lambda}{2} V |\varphi|^4 - \sum_{\vec{k}} E_{\vec{k}} \sinh^2 \alpha_{\vec{k}}, \qquad (1.90)$$

demonstrating that this state is the ground state of the Hamiltonian $H^{(2)}$. But the most important consequence of the elegant analysis of Bogoliubov is the discovery of "phonons". Indeed the spectrum of quantum fluctuations around the ground state $|\Omega\rangle$ for $\vec{k} \to 0$ vanishes linearly, and is characterized by a sound velocity ($E_{\vec{k}} \to v_s |\vec{k}|$ for $\vec{k} \to 0$)

$$v_s = |\varphi| \frac{\lambda^{1/2}}{m^{1/2}}. \qquad (1.91)$$

The normalization equation $\int_{\vec{x}} \Psi^\dagger \Psi = N$ determines $|\varphi|$ to be approximately

$$|\varphi|^2 = \frac{N}{V} = \rho. \qquad (1.92)$$

Finally by a simple exercise (Ex.$\langle 1.4 \rangle$) one can show that the ground state $\left| \omega_{\vec{k}} \right\rangle$ of the oscillator $A_{\vec{k}}$, for which $A_{\vec{k}} \left| \omega_{\vec{k}} \right\rangle = 0$, has in the Fock space the following form:

$$\left| \omega_{\vec{k}} \right\rangle = \frac{1}{\cosh \alpha_{\vec{k}}} \sum_{n=0}^{\infty} \frac{(-\tanh \alpha_{\vec{k}})^n}{n!} (a_{\vec{k}}^\dagger a_{-\vec{k}}^\dagger)^n \left| 0 \right\rangle, \qquad (1.93)$$

an infinite superposition of states containing increasing numbers of pairs of atoms with opposite momenta.

The simple and elegant analysis that we have just gone through has given substance to our expectation that coherence over long distances demands an interacting theory that cannot be treated perturbatively. What is, however, quite surprising is the fact that a short-range interaction, such as is described by the Hamiltonian (1.81), should give rise to coherence at any space-scale. Indeed, in a world governed by short-range interactions only we would expect such a coherence to disappear for large momenta $\vec{k}$, for in such systems the excitation of momenta $\vec{k}$ can be analysed as well in boxes of size $V_{\vec{k}} \sim (2\pi/|\vec{k}|)^3$, there existing no "fundamental" coherence length belonging to the interaction. If we do this, and "shrink" our box accordingly, we find that the last two terms in the quadratic Hamiltonian $H^{(2)}$ (1.87) cannot have sharp phase factors $e^{2i\chi}$ and $e^{-2i\chi}$ respectively, but they must be averaged in accordance with the indeterminacy relation (1.14)

$$\Delta\chi \sim \frac{1}{2\Delta N_{\vec{k}}} \simeq \frac{1}{2\sqrt{N_{\vec{k}}}} = \frac{1}{2} \left(\frac{|\vec{k}|}{2\pi} \right)^{3/2} (\rho)^{-1/2}. \qquad (1.94)$$

In this way, for $\vec{k}$'s such that $\Delta\chi \geq 1$ phase coherence is lost and the spectrum becomes

$$E_{\vec{k}} \simeq \left(\frac{\vec{k}^2}{2m} + \lambda \frac{N}{V}\right), \tag{1.95}$$

as one should have expected from a "mean field" approximation, which simply shifts the free spectrum by the mean interaction energy $\lambda(N/V)$. However such a loss of coherence at short distances, typical of short-range interactions, is not a general phenomenon of Condensed Matter physics. In the following Chapters it will become clear that the quantum electrodynamical interactions in Condensed Matter, characterized by intrinsic distance scales — the coherence domains — may lead to coherent behaviours completely different from the one discussed in this Section, which we may call "mean field" coherence. Indeed, by introducing the interaction between matter and the e.m. radiation field, a crucial missing element in the presently generally accepted condensed matter physics (GACMP), we will see that in particular conditions a non-trivial "classical" e.m. field arises in matter, that through the electrodynamics equations guarantees that the e.m. coherence is shared by the matter wave-fields as well.

Exercises of Chapter 1

$\langle 1.1\rangle$: Prove that from Eq. (1.12) there follow the fundamental phase-number commutation relations

$$[\theta_n, N_n] = -i.$$

Hint: Set $[\theta_n, N_n] = c$, expand $e^{i\theta_n} = \sum_k \frac{(i\theta_n)^k}{k!}$ and insert in $[e^{i\theta_n}, N_n] = e^{i\theta_n}$, which follows trivially from Eq. (1.12). This fixes the c-number $c = -i$.

$\langle 1.2\rangle$: From Eqs. (1.24) and (1.25) and definitions (1.32), derive (1.38) and (1.39).

$\langle 1.3\rangle$: Derive Eq. (1.55).
Hint: Use representation (1.52) and introduce for the complex variable α polar coordinates.

$\langle 1.4\rangle$: Derive Eq. (1.93).
Hint: Write $\left|\omega_{\vec{k}}\right\rangle$ as a superposition of n-pair states $(a_{\vec{k}}^+ a_{-\vec{k}}^+)^n \left|0\right\rangle$ with arbi-

trary n, and using the "annihilation" relation

$$A_{\bar{k}}\left|\omega_{\bar{k}}\right\rangle = (\cosh\alpha_k a_{\bar{k}} + \sinh\alpha_k a^+_{-\bar{k}})\left|\omega_{\bar{k}}\right\rangle = 0,$$

derive for the superposition coefficients c_n recursive relations.

References to Chapter 1

1. For a widely documented historical account of Planck's discovery see:
   T.S. Kuhn, *Black-body theory and the quantum discontinuity, 1894-1912* (Chicago Univ. Press, 1987).
2. A standard text where these notions are thoroughly discussed is:
   J.D. Bjorken and S.D.Drell, *Relativistic Quantum Fields*, (McGraw-Hill, New York, 1964).
3. The quantization of the free e.m. field that played such a fundamental rôle in the development of QFT appears very early in QM; the most important founding papers are:
   P.A.M. Dirac, *Proc. Roy. Soc.* **A114** (1927) 243, 710.
   P. Jordan and W. Pauli, *Zs. f. Phys.* **47** (1928) 151.
   W. Heisenberg and W. Pauli, *Zs. f. Phys.* **56** (1929) 1; **59** (1930) 169.
   E. Fermi, *Revs. Mod. Phys.* **4** (1931) 131.
4. This Section follows closely the scheme presented in the lectures: G. Preparata, *Quantum Field Theory of Superradiance*, in *Problems in Fundamental Modern Physics*, eds. R. Cherubini, P. Dalpiaz and B. Minetti (World Scientific, Singapore, 1990).
5. N.N. Bogoliubov, *On the theory of Superfluidity, Izv. Akad. Nauk. SSSR. Ser. Fiz.* **11** (1977) 77.

Chapter 2

THE DYNAMICS OF QED IN CONDENSED MATTER

2.1 Another formulation of the quantum kinematics of QFT: the path integral approach (PIA)

The invention of QM, in the middle of the twenties, required the great intellectual leap to think of the physical observables no more as real functions of sets of real variables (the fields) or as real variables themselves (the Lagrangian coordinates) but as linear operators defined in appropriate Hilbert spaces. In this way QM operated a sharp distinction between the notion of physical state (the vector of the Hilbert space) and that of physical observable, which on a given physical state needs not assume in general a well defined value. As we know, such a great departure from classical physics was forced by the necessity to incorporate in the wave concept the discontinuities of Planck's quanta, and in the particle concept the continuities of the Schrödinger wave-function. This wave-particle "duality", whose most significant expression is to be found in the Heisenberg indeterminacy relations, leads to the collapse of the notion of particle trajectories, due to the impossibility of sharply defining positions and velocities (momenta) simultaneously on a given quantum state, as quantified by the Heisenberg's principle itself.

The question of how and in what limit Classical Mechanics (CM) was to be related to Quantum Mechanics (QM) was only partly answered by Bohr's "correspondence principle", which requires the two mechanics to give the same answers in the limit of quantum states whose quantum numbers attain large values: a "principle" that, unlike the "complementarity principle", proved of great heuristic value. However if one wished to understand in detail how the two mechanics merged in the "correspondence" limit, Bohr's principle did not give any explicit clue. For that one had to wait until 1948 when Richard Feynman[1], taking up a luminous earlier (1937) suggestion by P.A.M. Dirac[2], solved the problem completely. And in so doing, he gave a new formulation of QM — the path integral approach (PIA) — that in QFT turns out to be extremely enlightening and useful.

Let us consider the simplest quantum mechanical system, described by a single Lagrangian coordinate q, with Lagrange-function $L(q, \dot{q}; t)$, conjugate momentum

$$p = \frac{\partial L}{\partial \dot{q}}, \tag{2.1}$$

and Hamiltonian-function

$$H(p, q) = p\dot{q} - L. \tag{2.2}$$

The object that the PIA wishes to describe in a new way is the transition amplitude $\langle q_f t_f \mid q_i t_i \rangle$ from an eigenstate $\mid q_i \rangle$ of the coordinate q at the initial time t_i to an eigenstate $\mid q_f \rangle$ at the final time $t_f > t_i$. In the usual formulation such a transition amplitude is given by (we reinstate, but only in this Section, the Planck constant $\hbar$)

$$\langle q_f t_f \mid q_i t_i \rangle = \langle q_f \mid e^{-i\frac{H}{\hbar}(t_f - t_i)} \mid q_i \rangle, \tag{2.3}$$

which follows from the Schrödinger equation.

The main idea of the PIA is to divide the time interval (t_i, t_f) in $\nu = \frac{t_f - t_i}{\Delta t}$ small intervals Δt with the aim to take the limit $\nu \to \infty$ at the end. In this way the time intervals get intercalated by the discrete times $t_j = t_i + j\Delta t$ $(j = 0, 1, ..., \nu)$, and the group properties of the time-evolution operator of QM allow us to write

$$\langle q_f t_f \mid q_i t_i \rangle = \int \prod_{j=0}^{\nu-1} \langle q_{j+1} \mid e^{-i\frac{H}{\hbar}\Delta t} \mid q_j \rangle \, dq_{j+1} \tag{2.4}$$

where the completeness of the states $\mid q \rangle$ has been used. Let us analyse the transition amplitudes $\langle q_{j+1} \mid e^{-i\frac{H}{\hbar}\Delta t} \mid q_j \rangle$ in an infinitesimal time-interval of length Δt. We have obviously

$$\langle q_{j+1} \mid e^{-i\frac{H}{\hbar}\Delta t} \mid q_j \rangle = \langle q_{j+1} \mid (1 - i\frac{H}{\hbar}\Delta t) \mid q_j \rangle$$

$$= \int dp_{j+1} \langle q_{j+1} \mid p_{j+1} \rangle \langle p_{j+1} \mid (1 - i\frac{H}{\hbar}\Delta t) \mid q_j \rangle$$

$$= \int \frac{dp_{j+1}}{2\pi\hbar} \exp \left\{ \frac{i}{\hbar}[(q_{j+1} - q_j)p_{j+1} - H(p_{j+1}, q_j)\Delta t] \right\} \tag{2.5}$$

where we have used the completeness of the momentum eigenstates $\mid p \rangle$, $\langle q \mid p \rangle = \frac{e^{-i\frac{pq}{\hbar}}}{\sqrt{2\pi\hbar}}$ as dictated by the canonical commutation relation $[p, q] = -i\hbar$, and in the expression of the Hamiltonian operator $H(q,p)$ we have adopted the ordering of the operators p and q for which all p's appear on the left of all q's. Note that we choose this ordering as the natural way to dispose of the intrinsic ambiguity that one encounters in translating a classical hamiltonian into a quantum mechanical operator. By inserting (2.5) in (2.4) we get:

$$\langle q_f t_f \mid q_i t_i \rangle = \int \prod_{j=0}^{\nu-1} \frac{(dp_{j+1} dq_{j+1})}{(2\pi\hbar)} \cdot \exp\left\{ \frac{i}{\hbar} \sum_{j=0}^{\nu-1} [\dot{q}_j p_{j+1} - H(p_{j+1}, q_j)] \Delta t \right\}. \qquad (2.6)$$

This integral representation of the quantum-mechanical transition amplitude is at this stage approximate (for we have not yet taken the continuum limit $\nu \to \infty$ ($\Delta t \to 0$)), but extremely interesting. The integration in (2.6) is over the classical phase space p-q for each discrete time, corresponding to summing over all phase space classical trajectories (paths!) sampled at discrete, but very close, times. The integrand can be easily recognized as the exponential of the discrete approximation to the classical action

$$A(t_i, t_f) = \int_{t_i}^{t_f} dt L(t) = \int_{t_i}^{t_f} dt[\dot{q}p - H] \qquad (2.7)$$

multiplied by the factor $\frac{i}{\hbar}$. The interpretation of (2.6) is now very simple: the quantum mechanical transition amplitude from a state $|q_i\rangle$ at the time t_i to a state $|q_f\rangle$ at the time t_f can be obtained by "summing" over all classical trajectories $(q(t), p(t))$ of the mechanical system, that begin in q_i and end in q_f, each trajectory being "weighted" by the phase-factor $\exp\left\{\frac{i}{\hbar} A(t_i, t_f)\right\}$, where $A(t_i, t_f)$ is the classical action belonging to that particular classical trajectory.

It is thus that the PIA dramatically reveals the differences between QM and CM: in the latter the transition from the initial to the final state involves only **one** path, the one for which the action is stationary, as implied by the Lagrange equations of motion. In the former, on the other hand, an infinite number of **classical** trajectories are sampled and superposed in the transition from the initial to the final state, their relative importance being determined by the behaviour of the **classical** action between neighbouring trajectories. The representation (2.6) contains in a very explicit fashion the physical quantity that demarcates classical from quantum behaviour, the Planck constant $\hbar$. When $\hbar \to 0$, or better when the classical actions involved are much larger than $\hbar$, it is a simple and well known mathematical result that the classical trajectories which contribute most of the path integral are those around the one which renders the action **stationary** , precisely the one which provides the solution of the dynamical problem of CM. For finite $\hbar$, the existence of many trajectories that link the initial to the final state, a fact that in QM leads to the dissolution of the notion of trajectory, can be given in the PIA a well defined quantitative meaning by observing that all paths will contribute to the transition whose actions differ from the classical action by $O(\hbar)$.

Taking now the continuum time limit $\Delta t \to 0$ ($\nu \to \infty$) (2.6) becomes

$$\langle q_f t_f \mid q_i t_i \rangle = \int \frac{[dp(t)dq(t)]}{(2\pi\hbar)} \exp\left\{ \frac{i}{\hbar} \int_{t_i}^{t_f} (\dot{q}p - H)dt \right\}$$

$$= \int \frac{[dp(t)dq(t)]}{(2\pi\hbar)} \exp\left\{ \frac{i}{\hbar} \int_{t_i}^{t_f} L(t)dt \right\}, \tag{2.8}$$

where $[dp(t)dq(t)]$ denotes the integration over an infinite number of p and q variables, one pair for each time $t \in (t_i, t_f)$. It is worth noting that, as a matter of fact, the integration over continuous functions, expressed by the functional integral (2.8), can be properly defined by the limiting procedure that leads us from (2.6) to (2.8). By introducing the oscillator amplitudes

$$a(t) = \frac{q(t) + ip(t)}{\sqrt{2}}, \tag{2.9}$$

$$a(t)^* = \frac{q(t) - ip(t)}{\sqrt{2}}, \tag{2.10}$$

classical counterparts of the quantum amplitudes obeying the standard commutation relations:

$$[a(t), a^\dagger(t)] = \hbar, \tag{2.11}$$

one can cast (2.8) in the form:

$$\langle q_f t_f \mid q_i t_i \rangle = \int \frac{[da(t)da(t)^*]}{(2\pi\hbar i)} \exp\left\{ \frac{i}{\hbar} \int_{t_i}^{t_f} L(t)dt \right\}. \tag{2.12}$$

Consider now the case of the anharmonic oscillator, whose Lagrangian is

$$L = \frac{\dot{q}^2}{2} - \frac{q^2}{2} + V(q^2), \tag{2.13}$$

$V(q^2)$ containing terms in q of order higher than q^2. From (2.8) and (2.10) we have

$$q = \frac{1}{\sqrt{2}}(a + a^*), \tag{2.14}$$

$$\dot{q} = \frac{1}{\sqrt{2}}(\dot{a} + \dot{a}^*). \tag{2.15}$$

Going to the "interaction" representation by the definitions

$$a = \alpha e^{-it}, \tag{2.16}$$

$$a^* = \alpha^* e^{it}, \tag{2.17}$$

the measure in (2.12) remains unchanged, while the action becomes

$$\int_{t_i}^{t_f} dt L = \int_{t_i}^{t_f} dt \left\{ \frac{1}{2}\dot{\alpha}^*\dot{\alpha} + \frac{i}{2}(\alpha^*\dot{\alpha} - \dot{\alpha}^*\alpha) + \frac{1}{4}e^{-2it}[(\dot{\alpha} - i\alpha)^2 - \alpha^2] \right.$$

$$\left. + \frac{1}{4}e^{2it}[(\dot{\alpha}^* + i\alpha^*)^2 - \alpha^{*2}] + V\left[\left(\frac{\alpha e^{-it} + \alpha^* e^{it}}{\sqrt{2}}\right)^2\right] \right\}$$

$$\simeq \int_{t_i}^{t_f} dt \left[\frac{1}{2}\dot{\alpha}^*\dot{\alpha} + \frac{i}{2}(\alpha^*\dot{\alpha} - \dot{\alpha}^*\alpha) + V(\alpha^*\alpha) \right], \tag{2.18}$$

where we have neglected the terms with the time exponentials, that for long time intervals $(t_f - t_i)$ will be always justified provided the important paths $(\alpha(t), \alpha^*(t))$ are such that their frequency spectrum does not include the frequencies $|\omega| = 1$ and $|\omega| = 2$.

The sense of these last developments can be appreciated by applying the PIA to the trivial, harmonic oscillator case $V(q^2) = 0$. One obtains [Ex. $\langle 2.1 \rangle$]

$$\langle q_f t_f \mid q_i t_i \rangle = \langle \alpha_f e^{-it_f}, t_f \mid \alpha_i e^{-it_i}, t_i \rangle$$

$$= \int \frac{[d\alpha(t)d\alpha^*(t)]}{(2\pi\hbar i)} \exp\left\{ \frac{i}{\hbar} \int dt[\frac{1}{2}\dot{\alpha}^*\dot{\alpha} + \frac{i}{2}(\alpha^*\dot{\alpha} - \dot{\alpha}^*\alpha)] \right\} \tag{2.19}$$

$$= \delta^{(2)}(\alpha_f - \alpha_i).$$

The generalization of the path-integral representation (2.8) to the case of systems with N Lagrangian degrees of freedom q_l $(l = 1, 2, ..., N)$ is completely straightforward:

$$\langle q_1^f...q_N^f, t_f \mid q_1^i...q_N^i, t_i \rangle = \int \prod_{l=1}^{N} \frac{[dp_l(t)dq_l(t)]}{(2\pi\hbar)} \exp\left\{ \frac{i}{\hbar} \int_{t_i}^{t_f} dt L(t) \right\}, \tag{2.20}$$

and likewise straightforward is the generalization to a quantum field system $\varphi(\vec{x}, t)$. Indeed the limiting process, that through the discretization of the time interval (t_i, t_f) led to the definition of the path-integral (2.8), can be generalized without any conceptual difficulties to the continuum limit of a space lattice, such as we have adopted in Chapter 1 as a means of defining a QFT. In the case of the QFT of a scalar field $\varphi(\vec{x}, t)$ the transition amplitude $\langle \varphi(\vec{x}, t_f) \mid \varphi(\vec{x}, t_i) \rangle$, from a "classical" field configuration $\varphi(\vec{x}, t_i)$ at the initial time t_i to a configuration $\varphi(\vec{x}, t_f)$ at the final time t_f, can be thus represented:

$$\langle \varphi(\vec{x}, t_f) \mid \varphi(\vec{x}, t_i) \rangle = \int \frac{[d\pi(\vec{x}, t) d\varphi(\vec{x}, t)]}{(2\pi\hbar)} \exp\left\{\frac{i}{\hbar} \int_{t_i}^{t_f} dt \int d^3\vec{x} \mathcal{L}(\vec{x}, t)\right\}, \qquad (2.21)$$

where $\pi(\vec{x}, t)$ is the classical momentum conjugate to the field $\varphi(\vec{x}, t)$

$$\pi(\vec{x}, t) = \frac{\partial \mathcal{L}}{\partial \dot{\varphi}(\vec{x}, t)}, \qquad (2.22)$$

and $\mathcal{L}(\vec{x}, t)$ is the Lagrangian density — function of $\varphi, \vec{\nabla}\varphi$ and $\dot{\varphi}$ — defining the QFT under consideration. The generalization of (2.21) to the case of many different bosonic fields $\varphi_k(\vec{x}, t)$ is totally obvious.

We end this Section by discussing the extension of the PIA to fermion degrees of freedom, whose quantum amplitudes $a(t)$ $(a^\dagger(t))$ instead of the standard commutation relation obey the anticommutation relation

$$[a(t), a^\dagger(t)]_+ = \hbar. \qquad (2.23)$$

In order to proceed we need introduce a "classical" fermionic phase space for the single degree of freedom, in analogy with what has been done for the bosonic phase space p-q or, equivalently, a-a^*. The basic difference between fermionic quantum oscillators, whose amplitudes obey (2.23), and their bosonic counterparts, obeying (2.11), is that their classical limits $\hbar \to 0$ in the latter case reduce to c-numbers, while in the former they constitute an algebraic system of anticommuting elements, that is called Grassmann algebra. A Grassmann algebra is thus a complex algebra of elements $a, b..., a^*, b^*, ...$, defined in a complex vector space, endowed with the multiplication operation and enjoying the distributive, associative and the anticommutative property:

$$ab + ba = 0 \qquad (2.24)$$

for any element $a, b, ...$

The analysis of complex functions defined on a Grassmann algebra follows the normal rules, provided one takes due account of ordering and anticommuting. As we know the quantum mechanical Hilbert space of the fermionic oscillator, obeying (2.23), is two-dimensional, consisting of the ground state $|0\rangle$, such that $a\,|0\rangle = 0$, and the excited state $|1\rangle = a^\dagger\,|0\rangle$, which due to the anticommutation relation is annihilated by $a^\dagger$. The generic state $|f\rangle$ can be thus represented as

$$|f\rangle = \alpha\,|0\rangle + \beta\,|1\rangle. \tag{2.25}$$

In order to give a treatment of the Fermi oscillator, parallel to the coherent state representation of the Bose oscillator, let us introduce the "wave-function" of the state $|f\rangle$ defined on the Grassmann algebra (for convenience we set now $\hbar = 1$)

$$|f\rangle \Rightarrow f(a) = \alpha a + \beta \tag{2.26}$$

such that

$$a\,|f\rangle \Rightarrow af(a) = \beta a, \tag{2.27}$$

and

$$a^\dagger\,|f\rangle \Rightarrow \frac{d}{da}f(a) = \alpha, \tag{2.28}$$

whose correctness can be readily checked.

The scalar product $\langle f'\,|\,f\rangle$ can be easily seen to be given by [Ex. $\langle 2.2\rangle$]

$$\langle f'\,|\,f\rangle = \alpha'^*\alpha + \beta'^*\beta = \int dada^* f'^*(a^*)K(a^*,a)f(a) \tag{2.29}$$

where $K(a^*,a) = e^{a^*a}$ provided one uses the following integration rules:

$$\int da = \int da^* = 0, \tag{2.30}$$

$$\int daa = \int da^*a^* = 1. \tag{2.31}$$

Discretizing as usual the time interval (t_i, t_f) we have

$$\langle f', t_f\,|\,f, t_i\rangle = \prod_{j=1}^{\nu-1}\left(\sum_{f_j}\langle f_{j+1}\,|\,e^{-iH\Delta t}\,|\,f_j\rangle\right)$$

$$= \int da_i da_f^* e^{a_f^* a_i} f^*(a_f) K(a_f, a_i^*; t_f, t_i) f(a_i), \qquad (2.32)$$

with

$$K(a_f, a_i^*; t_f, t_i) = \int \prod_{j=1}^{\nu-1} da_j da_j^* \exp\left\{ \sum_j \left[a_{j+1}^*(a_j - a_{j+1}) - iH(a_j^*, a_j)\Delta t \right] \right\}. \quad (2.33)$$

Taking the continuum limit we can finally write:

$$K(a_f, a_i^*; t_f, t_i) = \int [da(t) da^*(t)] \exp\left\{ i \int_{t_i}^{t_f} dt[ia^*(t)\dot{a}(t) - H(a^*, a)] \right\}, \qquad (2.34)$$

which is completely analogous to its bosonic counterparts, provided we interpret the expression $ia^*\dot{a} - H(a^*, a)$ as the Lagrange-function of the fermionic system, as we must obviously do.

The generalization of (2.34) to a fermionic quantum field is again straightforward.

2.2 "Condensation" and the large-N limit of QED in Condensed Matter

We possess now all the necessary elements to begin a general analysis of the dynamics of the QED of a matter system described by the quantum field $\Psi(\vec{x}, \alpha, t)$. Following the developments of Chapter 1 the Lagrange function of the matter system, which for the time being is taken to be bosonic, is

$$L_{matt} = i \int_{\vec{x}\alpha} \Psi^\dagger(\vec{x}, \alpha, t) \frac{\partial}{\partial t} \Psi(\vec{x}, \alpha, t) - H_{matt}, \qquad (2.35)$$

$$H_{matt} = \int_{\vec{x}\alpha} \Psi^\dagger(\vec{x}, \alpha, t) H_o(\vec{x}, \alpha) \Psi(\vec{x}, \alpha, t) + H_{rad}^{(1)} + H_{rad}^{(2)} + H_{SR}. \qquad (2.36)$$

As for the transverse electromagnetic field, which we write in the plane-wave expansion [see (1.32a)]

$$\vec{A}(\vec{x}, t) = \sum_{\vec{k}r} \frac{1}{\sqrt{2\omega_{\vec{k}} V}} \left[a_{\vec{k}r}(t) \vec{\epsilon}_{\vec{k}r} e^{-i\omega_{\vec{k}} t} e^{i\vec{k}\cdot\vec{x}} + a_{\vec{k}r}(t)^* \vec{\epsilon}_{\vec{k}r}^* e^{i\omega_{\vec{k}} t} e^{-i\vec{k}\cdot\vec{x}} \right] \qquad (2.37)$$

where $\omega_{\vec{k}} = |\vec{k}|$, $\vec{k} \cdot \vec{\epsilon}_{\vec{k}r} = 0$ and $\vec{\epsilon}_{\vec{k}r} \cdot \vec{\epsilon}_{\vec{k}s} = \delta_{r,s}$, we have seen in Chapter 1 that it can be looked upon as a collection of "ether oscillator", one for each plane-wave mode $(\vec{k}, r)$. Thus its Lagrangian L_{em}, to be used in the PIA, is given by the sum over all such modes of expressions of the type (2.18) yielding:

$$L_{em} = \sum_{\vec{k}r} \left[\frac{i}{2}(a_{\vec{k}r}(t)^* \dot{a}_{\vec{k}r}(t) - \dot{a}_{\vec{k}r}(t)^* a_{\vec{k}r}(t)) + \frac{1}{2\omega_{\vec{k}}} \dot{a}_{\vec{k}r}(t)^* \dot{a}_{\vec{k}r}(t) \right]. \tag{2.38a}$$

The e.m. hamiltonian $H_{em} = \frac{1}{2}\int_{\vec{x}} \frac{(\vec{E}^2 + \vec{B}^2)}{2}$ is easily shown to equal [Ex. ⟨2.3⟩]:

$$H_{em} = \sum_{\vec{k}r} \left[\omega_{\vec{k}} a_{\vec{k}r}(t)^* a_{\vec{k}r}(t) + \frac{i}{2}(a_{\vec{k}r}(t)^* \dot{a}_{\vec{k}r}(t) - \dot{a}_{\vec{k}r}(t)^* a_{\vec{k}r}(t)) + \frac{1}{2\omega_{\vec{k}}} \dot{a}_{\vec{k}r}(t)^* \dot{a}_{\vec{k}r}(t) \right]$$

$$\tag{2.38b}$$

We can now write down the path-integral representation for the transition amplitude

$$\langle f, t_f \mid i, t_i \rangle_N = \int [d\Psi^\dagger(\vec{x}, \alpha, t)][d\Psi(\vec{x}, \alpha, t)][da_{\vec{k}r}(t)][da_{\vec{k}r}(t)^*] e^{\frac{i}{\hbar}\int_{t_i}^{t_f} dt(L_{matt} + L_{em})}, \tag{2.39}$$

where the suffix N alludes to the constraint that we are integrating over all paths $\{\Psi, \Psi^\dagger, a_{\vec{k}r}, a_{\vec{k}r}^\dagger\}$ for which the number of matter-systems in the interaction volume V is fixed and equal to N: on such paths

$$\int_{\vec{x}\alpha} \Psi^\dagger(\vec{x}, \alpha, t)\Psi(\vec{x}, \alpha, t) = N \tag{2.40}$$

holds. The existence of a large number N suggests rescaling the theory so as to make N appear explicitly in the path-integral. We thus define

$$\Psi_o(\vec{x}, \alpha, t) = \frac{1}{\sqrt{N}}\Psi(\vec{x}, \alpha, t), \tag{2.41}$$

and

$$a_{\vec{k}r}^o = \frac{1}{\sqrt{N}}a_{\vec{k}r}, \tag{2.42}$$

so that (2.40) becomes:

$$\int_{\vec{x}\alpha} \Psi_o(\vec{x}, \alpha, t)^\dagger \Psi_o(\vec{x}, \alpha, t) = 1. \tag{2.43}$$

We can rewrite the path-integral representation (2.39)

$$\langle f, t_f \mid i, t_i \rangle_N \propto \int [d\Psi_o][d\Psi_o^\dagger][da_{\vec{k}r}^o][da_{\vec{k}r}^{o\dagger}] \exp\left\{ i\frac{N}{\hbar}\int_{t_i}^{t_f} dt(\bar{L}_{matt} + \bar{L}_{em}) \right\}, \tag{2.44}$$

where

$$\bar{L}_{matt} = L_{matt}[\Psi_o, \Psi_o^\dagger; a_{\vec{k}r}^o, a_{\vec{k}r}^{o\dagger}; \sqrt{N}e],\tag{2.45}$$

and

$$\bar{L}_{em} = L_{em}[a_{\vec{k}r}^o, a_{\vec{k}r}^{o\dagger}].\tag{2.46}$$

The simple rescalings (2.41) and (2.42) have produced a remarkable and very suggestive result. In terms of the rescaled fields ψ_o and $a_{\vec{k}r}^o$, the path-integral retains its original form with two very important differences:

(i) the explicit factorization of the large number N, multiplying an action that in view of the normalization (2.43) is bounded from below;

(ii) the effective electromagnetic coupling e gets amplified by $\sqrt{N}$, suggesting a basically collective character of the interaction between the radiative e.m. field and matter.

It is just for this latter aspect that I thought it appropriate to call the dynamics implied by (2.44), (2.45) and (2.46) **superradiant**, thereby paying a tribute to the seminal early work of R.H. Dicke[3]. Having perceived with great surprise that this tribute created confusion and antagonism in the scientific community, I find myself in the disagreeable necessity to drop this term and simply commit it to the history of the development of the physical ideas expounded in this book.

The most significant aspect of large N's, such as one finds in condensed matter, is their playing a rôle in the PIA completely equivalent to $\hbar \to 0$, thereby expressing a dynamics whose structure is basically "classical". Its fully quantum character, however, must be appreciated with no hesitation, for Planck's constant remains finite (and indeed $\hbar = 1$ shall be resumed from now on), and most importantly the matter wave-field, whose "classical" limit will provide a kind of "macroscopic" Schrödinger wave-function, is all but a classical concept.

The theorem of the stationary phase as applied to the integral (2.44) in the large N-limit implies that the matter wave-field Ψ_o and the e.m. field amplitudes $a_{\vec{k}r}^o$ can be written as

$$\Psi_o(\vec{x}, \alpha, t) = \varphi(\vec{x}, \alpha, t) + \frac{1}{\sqrt{N}}\eta(\vec{x}, \alpha, t)\tag{2.47}$$

$$a_{\vec{k}r}^o(t) = \alpha_{\vec{k}r}(t) + \frac{1}{\sqrt{N}}\beta_{\vec{k}r}(t),\tag{2.48}$$

where $\varphi(\vec{x}, \alpha, t)$ and $\alpha_{\vec{k}r}(t)$ are c-number functions, and the quantum field $\eta(\vec{x}, \alpha, t)$ and the amplitudes $\beta_{\vec{k}r}(t)$ represent quantum fluctuations of order $\frac{1}{\sqrt{N}}$ around the classical state whose path is described by $\varphi(\vec{x}, \alpha, t)$ and $\alpha_{\vec{k}r}(t)$. Furthermore the classical path is completely determined by the variational principle

$$\delta \int_{t_i}^{t_f} (\bar{L}_{matt} + \bar{L}_{em}) = 0, \qquad (2.49)$$

i.e. the principle of extremal action.

By varying the action with respect to $\Psi(\vec{x}, \alpha, t)^*$ and neglecting for the time being H_{SR}, we readily obtain:

$$i\frac{\partial}{\partial t}\varphi(\vec{x}, \alpha, t) = H_o(\vec{x}, t)\varphi(\vec{x}, \alpha, t) + e\sqrt{N}\vec{A}_o(\vec{x}, t) \cdot \vec{J}(\alpha)\varphi(\vec{x}, \alpha, t); \qquad (2.50)$$

while by varying with respect to $\alpha_{\vec{k}r}^*$ we derive at once:

$$-\frac{1}{2\omega_{\vec{k}}}\ddot{\alpha}_{\vec{k}r}(t) + i\dot{\alpha}_{\vec{k}r}(t) - \frac{e^2}{\omega_{\vec{k}}}\left(\frac{N}{V}\right)\lambda\alpha_{\vec{k}r}(t)$$

$$= e\frac{1}{\sqrt{2\omega_{\vec{k}}}}\left(\frac{N}{V}\right)^{1/2}\vec{\epsilon}_{\vec{k}r}^{*}e^{i\omega_{\vec{k}}t}\int_{\vec{x}\alpha}e^{-i\vec{k}\cdot\vec{x}}\varphi(\vec{x}, \alpha, t)^*\vec{J}(\alpha)\varphi(\vec{x}, \alpha, t). \qquad (2.51)$$

In the expression for L_{matt} we have simplified the term $H_{rad}^{(2)}$ as follows:

$$H_{rad}^{(2)} = e^2\lambda \int_{\vec{x}\alpha} \vec{A}(\vec{x}, t)^2 \Psi^\dagger(\vec{x}, \alpha, t)\Psi(\vec{x}, \alpha, t)$$

$$\simeq e^2\lambda\left(\frac{N}{V}\right)\int_{\vec{x}} \vec{A}_o(\vec{x}, t)^2 = e^2\left(\frac{N}{V}\right)\sum_{\vec{k}r}\frac{\lambda}{\omega_{\vec{k}}}\alpha_{\vec{k}r}^*\alpha_{\vec{k}r}, \qquad (2.52)$$

due to the expected slow space variation of the density operator $\int_\alpha \Psi^\dagger(\vec{x}, \alpha)\Psi(\vec{x}, \alpha) \simeq \left(\frac{N}{V}\right)$, and the negligible contribution to the action of the "time-rotating terms" $a_{\vec{k}r}a_{\vec{k}r}e^{-2i\omega_{\vec{k}}t}$ and $a_{\vec{k}r}^*a_{\vec{k}r}^*e^{2i\omega_{\vec{k}}t}$.

The structure of the evolution equations (that we shall call from now on "coherence equations" (CE's) in order to stress the coherent character of their solutions) determining the "classical path" is remarkable indeed: equation (2.50) is a kind of Schrödinger equation for the "wave-function" $\varphi(\vec{x}, \alpha, t)$ coupled through a "coherent" charge $\sqrt{N}e$ to the electromagnetic field $\vec{A}_o(\vec{x}, t)$, whose expression is identical to the standard $\vec{A}(\vec{x}, t)$ in (2.37) but for the substitution of $a_{\vec{k}r}$ $(a_{\vec{k}r}^*)$ with $\alpha_{\vec{k}r}$ $(\alpha_{\vec{k}r}^*)$ [see (2.42)]. The usual Schrödinger equation for a single system would have exactly the same form with the Schrödinger wave-function in the place of the classical wave-field $\varphi(\vec{x}, \alpha, t)$, which is normalized in the same way [see (2.43)], and the e.m. coupling e instead of $\sqrt{N}e$. As for Eq. (2.51) it is nothing else than the classical Maxwell equation in a medium with dielectric constant $\epsilon\omega = 1 - e^2 \frac{\lambda}{\omega^2} \left(\frac{N}{V}\right)$ whose source is the classical current associated with the matter system, whose coupling is again amplified by the coherence factor $\sqrt{N}$.

In view of the very strong coupling between the matter and the e.m. wave-fields implied by the dynamical equations (2.50) and (2.51) I believe that it should not be too difficult to agree on the rather low likelihood of the physics situation in which the two systems would in general experience only a perturbative coupling, as held true in GACMP. We thus expect that, even if the initial state corresponds to a "perturbative" configuration in which the "ether oscillators" perform their zero-point oscillations ($|\alpha_{\vec{k}r}| \simeq \frac{1}{\sqrt{N}}$), as a result of their mutual strong coupling the matter and e.m. systems will develop large $(O(1))$ e.m. amplitudes and currents: a phenomenon that is quite close to the "condensation" of an ordered fluid out of a "gas" of independent and disordered quantum fluctuations, in which the matter systems perform collective oscillations in phase with a peculiar "condensate" (coherent state) of the e.m. field.

To end this Section let us extend the fundamental dynamical equations (2.50) and (2.51) to the case of fermionic matter. Calling $|\Omega, t\rangle$ the state of the matter wave-field $\Psi(\vec{x}, \alpha, t)$ that at the time t evolves from the initial state $|\Omega\rangle$, the Schrödinger equation can be written:

$$i\frac{\partial}{\partial t} |\Omega, t\rangle = \int_{\vec{x}\alpha} \Psi^\dagger(\vec{x}, \alpha, t) H_o(\vec{x}, \alpha) \Psi(\vec{x}, \alpha, t) |\Omega, t\rangle$$

$$+ e \int_{\vec{x}\alpha} \vec{A}(\vec{x}, t) \Psi^\dagger(\vec{x}, \alpha, t) \vec{J}(\alpha) \Psi(\vec{x}, \alpha, t) |\Omega, t\rangle \,, \qquad (2.53)$$

while the Maxwell equation takes the form:

$$-\frac{1}{2\omega_{\vec{k}}}\ddot{\alpha}_{\vec{k}r}(t) + i\dot{\alpha}_{\vec{k}r}(t) - \frac{e^2}{\omega_{\vec{k}}} \left(\frac{N}{V}\right) \lambda\alpha_{\vec{k}r}(t)$$

$$= e\frac{1}{\sqrt{2\omega_{\vec{k}}V}} \vec{\epsilon}_{\vec{k}r} e^{i\omega_{\vec{k}}t} \int_{\vec{x}\alpha} e^{-i\vec{k}\cdot\vec{x}} \langle\Omega, t| \Psi^\dagger(\vec{x}, \alpha, t) \vec{J}(\alpha) \Psi(\vec{x}, \alpha, t) |\Omega, t\rangle \,. \qquad (2.54)$$

We shall see later how and in what situations these equations will reduce to a form very close to that of the Bose case. Here we note that the Schrödinger equation (2.53) is just the general dynamical equation for the evolution of the matter-field, and the Maxwell equation is just the "classical" equation for the e.m. field whose source is the classical e.m. current in the matter state $|\Omega, t\rangle$.

2.3 The dynamics of quantum fluctuations

Having determined the "path" $\varphi(\vec{x}, \alpha, t)$ and $\alpha_{\vec{k}r}(t)$ that makes the action

$$A[\psi, \psi^\dagger, a_{\vec{k}r}, a_{\vec{k}r}^\dagger] = \int_{t_i}^{t_f} dt (\bar{L}_{matt} + \bar{L}_{em}) \qquad (2.55)$$

stationary, we can now analyse the structure of the action in terms of the fluctuating field $\eta(\vec{x}, \alpha, t)$ and the amplitudes $\beta_{\vec{k}r}(t)$ and $\beta_{\vec{k}r}^\dagger(t)$ introduced in (2.47) and (2.48). By making such expansion around the "stationary path" one gets

$$A[\psi, \psi^\dagger, a_{\vec{k}r}, a_{\vec{k}r}^\dagger] = A[\phi, \phi^*, \alpha_{\vec{k}r}, \alpha_{\vec{k}r}^*] + \int \frac{\delta^2 A}{\delta \psi^\dagger \delta \psi} [\phi, \phi^*, \alpha_{\vec{k}r}, \alpha_{\vec{k}r}^*] \frac{1}{N} \eta^\dagger \eta$$

$$+ \sum_{\vec{k}r} \sum_{\vec{k}'s} \frac{\delta^2 A}{\delta a_{\vec{k}r}^\dagger \delta a_{\vec{k}'s}} [\phi, \phi^*, \alpha_{\vec{k}r}, \alpha_{\vec{k}r}^*] \frac{1}{N} \beta_{\vec{k}r}^\dagger \beta_{\vec{k}r} +$$

$$+ \text{(higher order terms including the short-range interactions)}, \qquad (2.56)$$

where evidently we have neglected the first order functional derivatives as they vanish due to the stationarity of the action on the "classical path" $\{\phi, \phi^*, \alpha_{\vec{k}r}, \alpha_{\vec{k}r}^*\}$. A very simple calculation [Ex. $\langle 2.4\rangle$] shows that the Lagrangian L_{fluc} governing the dynamics of the quantum fluctuations $\eta(\eta^\dagger), \beta_{\vec{k}r}(\beta_{\vec{k}r}^\dagger)$ is

$$L_{fluc} = \int_{\vec{x}\alpha} \eta(\vec{x}, \alpha, t)^\dagger (i\frac{\partial}{\partial t} - H_o(\vec{x}, \alpha))\eta(\vec{x}, \alpha, t)$$

$$+ e\sqrt{N} \int_{\vec{x}\alpha} \vec{A}_o(\vec{x}, t)\eta(\vec{x}, \alpha, t)^\dagger \vec{J}(\alpha)\eta(\vec{x}, \alpha, t)$$

$$+ \sum_{\vec{k}r} \left[\frac{i}{2}(\beta_{\vec{k}r}^\dagger(t)\dot{\beta}_{\vec{k}r}(t) - \dot{\beta}_{\vec{k}r}^\dagger(t)\beta_{\vec{k}r}(t)) + \frac{1}{2\omega_{\vec{k}}}\dot{\beta}_{\vec{k}r}^\dagger(t)\dot{\beta}_{\vec{k}r}(t) - \frac{e^2}{\omega_{\vec{k}}}\left(\frac{N}{V}\right)\lambda\beta_{\vec{k}r}^\dagger(t)\beta_{\vec{k}r}(t) \right],$$
$$(2.57)$$

indicating that the quantum fluctuations around both the matter and the e.m. "classical" fields correspond in the case of the matter field $(\eta, \eta^\dagger)$ to independent fluctuations coupled to the external e.m. field $\vec{A}_o(\vec{x}, t)$, given by the e.m. "condensate"; while in the case of the e.m. field $(\beta_{\vec{k}r}, \beta_{\vec{k}r}^\dagger)$ they differ from the free space e.m. waves by the "mass-term" $e^2\lambda\left(\frac{N}{V}\right)$, arising from the "dispersive" interaction of the e.m. field with matter.

Once due account is taken of the effect of the "external" field $\vec{A}_o(\vec{x}, t)$ and of the photon "mass-term" $e^2 \lambda \left(\frac{N}{V} \right)$, the physics of fluctuations becomes "perturbative", due to the general weakness of the residual interactions. Thus the dynamics of quantum fluctuations can be calculated systematically for the perfectly well defined "free" (independent) systems, through a straightforward perturbative expansion. Using the language of QFT we may conclude that perturbation theory gives an adequate description of the physics of quantum fluctuations, however it turns out that the physical ground state is not the perturbative (zero-condensate) one, but it has a highly non-trivial structure given by the solution of the "coherence equations" (2.50) and (2.51).

2.4   Temperature and the two-fluid picture

The dynamics of the QFT 's, that we have explored so far, makes no reference whatsoever to the notion of temperature, which is indeed totally extraneous to the deterministic evolution induced by the "coherence equations" (CE). In fact the physics of both matter and e.m. field that we have discussed so far refers to temperature T=0. Condensed matter systems at T=0, strictly speaking, do not exist and any realistic (hence controllable) physical description must provide some very definite idea on how temperature affects the "cold" dynamics that up to this point we have been exclusively involved in.

Let us look at our physical system at T=0. If the system is open its dynamical evolution, governed by the CE's, will lead it after a certain time to its ground state, i.e. to the state of minimum energy, which (as we shall always assume) is a particular solution of the CE's, as we shall see in the next Chapter. Around this ground state the system can fluctuate quantum mechanically and, from the discussion in the preceding Section, the quantum fluctuations can be thought of as comprising a "gas" of quasi-particles, matter and "photons", with a spectrum whose structure depends in general on the CE's. If we now let $T \neq 0$, Nernst's heat theorem will not guarantee any more that the system is all condensed in the ground state, and some "evaporation" will begin to take place, subtracting systems from the ground state and bringing them to populate the levels of the gas of quasi-particles. In this way the system separates in two distinct fluids, one completely ordered comprising the particles that move in phase with the e.m. field making up the CGS — we call it the "superfluid" —, the other — the "normal" fluid — composed by the gas of quasi-particles, with energy levels E_k, "bubbling" in the superfluid. How is this gas populated? From statistical thermodynamics we know that the occupation numbers n_k of the quasi-particle levels are Boltzmann distributed, i.e. (recall that we choose our units so that $k_B = 1$)

$$n_k \propto e^{-\frac{E_k}{T}}, \tag{2.58}$$

so that the overall population of the normal fluid is (assuming each level to be non-degenerate)

$$N_n \propto \sum_k e^{-\frac{E_k}{T}}, \tag{2.59}$$

and the population of the superfluid N_s is obtained from the conservation of the particle number operator:

$$N_s = N - N_n \tag{2.60}$$

When T increases N_n increases, while N_s, according to (2.60), decreases. This fact has a twofold effect on our system: it changes the properties of the ground state, which in their turn modify the spectrum of the normal fluid. In particular the energy "gap" between the superfluid and the normal state, that we expect to be generated by the energy minimization occurring in the ground state, with the decreasing of N_s will tend to decrease: we are bound for a "phase-transition". When will it occur ?

Quite generally we can envisage two distinct situations. In the first the density of our system varies little at the phase-transition temperature T^*, this means that such transition takes place simply by the complete "invasion" of the superfluid by the normal fluid, the e.m. condensate as well as the energy "gap" go continuously to zero at T^*; we have a "second order" phase-transition. In the second, the process of "invasion" of the superfluid by the normal fluid terminates abruptly, leading to an "explosion" of the normal fluid into a state whose spatial structure differs significantly from the one it had below T^*: this phase-transition is "first order". But more of this in the following Chapters.

Exercises of Chapter 2

⟨**2.1**⟩: Derive Eq. (2.19).

Hint: Integrate the action by parts to obtain

$$\int_{t_i}^{t_f} dt\, L(t) = \int_{t_i}^{t_f} dt\, \alpha^*(t) \left[-\frac{1}{2}\frac{d^2}{dt^2} + i\frac{d}{dt} \right] \alpha(t) + \frac{1}{2}\left[\alpha_f^*(\dot{\alpha}_f - i\alpha_f) - \alpha_i^*(\dot{\alpha}_i - i\alpha_i) \right].$$

Integrate over $\alpha^*(t)$ to get an infinite product of δ-functions. The δ-functions constrain to zero the amplitudes of all modes but the zero mode $\alpha(t) = \alpha_o$

(the mode $\alpha(t) = \alpha_2 e^{-2it}$ is excluded by our approximation). The integration over the zero mode amplitude α_o can be carried out directly with the phase $\exp\left\{\frac{i}{\hbar} A[\alpha_o, \alpha_o^*]\right\}$.

$\langle\mathbf{2.2}\rangle$: Derive Eq. (2.29) and (2.33).

Hint: Note that the completeness sum is given by

$$\sum_f |f\rangle\langle f| = aa^* + 1 = -a^*a + 1 = \exp(-a^*a).$$

$\langle\mathbf{2.3}\rangle$: Derive Eq. (2.38b).

$\langle\mathbf{2.4}\rangle$: Derive Eq. (2.57).

References to Chapter 2

1. R.P. Feynman, *Revs. Mod. Phys.* **20** (1948) 267.
2. P.A. Dirac, *Phys. Zeit. Sov. Un.* **3** (1933) 1.
3. R.H. Dicke, *Phys. Rev.* **93** (1954) 99.

Chapter 3

QED OF TWO-LEVEL SYSTEMS

3.1 The classical equations for two-level systems coupled to a single e.m. mode. The emergence of Coherence Domains

It seems a good strategy to begin our analysis of the CE's with the simplest of all matter systems: the two-level system.

The internal variables α of the wave-field $\Psi(\vec{x}, \alpha, t)$ in this case simply reduce to a single dichotomic variable $\alpha = 1, 2$, denoting the two relevant quantum states of an atom or molecule. Even though the two-level model may seem much too simplified, we must recall that much of laser physics is based upon this simple model, which as a matter of fact has come to play a sort of paradigmatic rôle. The reason for the importance, and indeed for the real physical relevance of such model is the selective nature of its coupling to the e.m. radiation field: calling $\omega_0 = E_1 - E_2$ the energy difference between the two atomic levels, the long-term dynamical evolution is easily seen to sharply enhance the amplitudes of those e.m. field modes that are in "resonance" with the two-level transition, for which evidently $\omega_{\vec{k}} = \omega_0$. And if for some reason a frequency lock-in situation establishes itself, it is clear that neglecting the many other atomic levels will prove totally inconsequential. This much for the motivation of the physical relevance of our system.

Let us now write down the CE's for the "classical" limit $\varphi_i(\vec{x}, t)$ of the wave-field $\psi_i(\vec{x}, t)$, and for the e.m. amplitudes field $\alpha_{\vec{k}r}$. One has (we set for simplicity $E_2 = 0$, and m denotes the mass of the atomic system)

$$i\frac{\partial}{\partial t}\varphi_2(\vec{x}, t) = -\frac{\vec{\nabla}^2}{2m}\varphi_2(\vec{x}, t) + eJ\sqrt{\frac{N}{V}}\sum_{\vec{k}r}\frac{1}{\sqrt{2\omega_{\vec{k}}}}\left[\alpha_{\vec{k}r}\epsilon_{\vec{k}r,1}e^{-i(\omega_{\vec{k}}t - \vec{k}\cdot\vec{x})} + \text{c.c.}\right]\varphi_1(\vec{x}, t),$$

$$(3.1a)$$

$$i\frac{\partial}{\partial t}\varphi_1(\vec{x},t) = \left(-\frac{\vec{\nabla}^2}{2m}+\omega_0\right)\varphi_1(\vec{x},t)$$

$$+eJ\sqrt{\frac{N}{V}}\sum_{\vec{k}r}\frac{1}{\sqrt{2\omega_{\vec{k}}}}\left[\alpha_{\vec{k}r}\epsilon_{\vec{k}r,1}e^{-i(\omega_{\vec{k}}t-\vec{k}\cdot\vec{x})}+\text{c.c.}\right]\varphi_2(\vec{x},t),$$

(3.1b)

$$-\frac{1}{2\omega_{\vec{k}}}\ddot{\alpha}_{\vec{k}r}+i\dot{\alpha}_{\vec{k}r}-\frac{e^2}{\omega_{\vec{k}}}\left(\frac{N}{V}\right)\lambda\alpha_{\vec{k}r}=eJ\sqrt{\frac{N}{V}}\frac{1}{\sqrt{2\omega_{\vec{k}}}}\epsilon_{\vec{k}r,1}^{*}e^{i\omega_{\vec{k}}t}\int_{\vec{x}}e^{-i\vec{k}\cdot\vec{x}}\left(\varphi_1^{*}\varphi_2+\varphi_2^{*}\varphi_1\right),$$

(3.1c)

where the current operator has been conventionally taken as $(J)_i = \delta_{i1}J\sigma_1$. We go now to the "interaction representation" defining

$$\varphi_i(\vec{x},t) = e^{-iE_i t}\chi_i(\vec{x},t),$$

(3.2)

this modifies the CE's (3.1) as:

$$i\frac{\partial}{\partial t}\chi_2(\vec{x},t) = -\frac{\vec{\nabla}^2}{2m}\chi_2(\vec{x},t)+eJ\sqrt{\frac{N}{V}}\sum_{\vec{k}r}\frac{1}{\sqrt{2\omega_{\vec{k}}}}\left[\alpha_{\vec{k}r}\epsilon_{\vec{k}r,1}e^{-i(\omega_{\vec{k}}t-\vec{k}\cdot\vec{x})}+\text{c.c.}\right]e^{-i\omega_0 t}\chi_1(\vec{x},t),$$

(3.3a)

$$i\frac{\partial}{\partial t}\chi_1(\vec{x},t) = -\frac{\vec{\nabla}^2}{2m}\chi_1(\vec{x},t)+eJ\sqrt{\frac{N}{V}}\sum_{\vec{k}r}\frac{1}{\sqrt{2\omega_{\vec{k}}}}\left[\alpha_{\vec{k}r}\epsilon_{\vec{k}r,1}e^{-i(\omega_{\vec{k}}t-\vec{k}\cdot\vec{x})}+\text{c.c.}\right]e^{i\omega_0 t}\chi_2(\vec{x},t),$$

(3.3b)

$$-\frac{1}{2\omega_{\vec{k}}}\ddot{\alpha}_{\vec{k}r}+i\dot{\alpha}_{\vec{k}r}-\frac{e^2}{\omega_{\vec{k}}}\left(\frac{N}{V}\right)\lambda\alpha_{\vec{k}r}$$

$$=eJ\sqrt{\frac{N}{V}}\frac{1}{\sqrt{2\omega_{\vec{k}}}}\epsilon_{\vec{k}r,1}^{*}e^{i\omega_{\vec{k}}t}\int_{\vec{x}}e^{-i\vec{k}\cdot\vec{x}}\left(\chi_1^{*}\chi_2 e^{i\omega_0 t}+\chi_2^{*}\chi_1 e^{-i\omega_0 t}\right).$$

(3.3c)

As we are interested in the long-time dynamics of the CE's, we are completely justified in neglecting all terms that explicitly contain time-oscillating exponential factors $e^{i\alpha t}$ that get averaged out for times $t > \frac{2\pi}{\alpha}$. In laser physics this neglect corresponds to the well-known "rotating-wave" approximation, which leads to the following enormous simplifications:

$$i\frac{\partial}{\partial t}\chi_2(\vec{x},t) = -\frac{\vec{\nabla}^2}{2m}\chi_2(\vec{x},t)+eJ\sqrt{\frac{N}{V}}\frac{1}{\sqrt{2\omega_0}}\sum_{|\vec{k}|=\omega_0,r}\vec{\epsilon}_{\vec{k}r}^{*}\alpha_{\vec{k}r}^{*}e^{-i\vec{k}\cdot\vec{x}}\chi_1(\vec{x},t),$$

(3.4a)

$$i\frac{\partial}{\partial t}\chi_1(\vec{x},t) = -\frac{\vec{\nabla}^2}{2m}\chi_1(\vec{x},t) + eJ\sqrt{\frac{N}{V}}\frac{1}{\sqrt{2\omega_0}}\sum_{|\vec{k}|=\omega_0,r}\vec{\epsilon}_{\vec{k}r}\alpha_{\vec{k}r}e^{i\vec{k}\cdot\vec{x}}\chi_2(\vec{x},t), \qquad (3.4b)$$

$$-\frac{1}{2\omega_0}\ddot{\alpha}_{\vec{k}r} + i\dot{\alpha}_{\vec{k}r} - \frac{e^2}{\omega_0}\left(\frac{N}{V}\right)\lambda\alpha_{\vec{k}r} = eJ\sqrt{\frac{N}{V}}\frac{1}{\sqrt{2\omega_0}}\epsilon_{\vec{k}r,1}^*\int_{\vec{x}}e^{-i\vec{k}\cdot\vec{x}}\chi_2^*\chi_1, \qquad (3.4c)$$

$$\alpha_{\vec{k}r} = 0 \qquad \text{for} \qquad |\vec{k}| \neq \omega_0. \qquad (3.4d)$$

Before analyzing in detail this differential system, there are a few important general observations that one can make.

First [see (3.4d)] the e.m. field remains in its perturbative ground state for all modes $\vec{k}, r$ that do not "resonate" with the intrinsic atomic oscillation ω_0. On the other hand the resonating modes, whose "vacuum" frequencies $\omega_{\vec{k}} = |\vec{k}|$ equal ω_0, turn out to be strongly coupled with the matter field χ_i provided the density $\left(\frac{N}{V}\right)$ is large enough. The selection of the modes for which $|\vec{k}| = \omega_0$, operated by the "rotating-wave" approximation, yields the remarkable result that, should the e.m. field evolve toward a non trivial "condensate", this latter will necessarily exhibit a space-structure comprising an array of **COHERENCE DOMAINS (CD)**, within which the "classical" e.m. field is coherent and varies very little. The size of such CD's is clearly of the order of the wave length $\lambda = \frac{2\pi}{\omega_0}$ of the selected e.m. modes, and inside them also the matter wave-functions $\chi_i(\vec{x},t)$ vary very slowly. In this way the volume V, occupied by the N two-level atoms, is broken up in a large array of CD's, within which the dynamical evolution of matter and of the e.m. field is **coherent** and **homogeneous**. The CD is thus the **smallest** spatial region where a coherent evolution of matter and e.m. field can take place. The aspect of the CD that seems most remarkable is the translation, that gets realized within it, of a particular frequency ω_0, belonging to the dynamics of the single atom, into a spatial modulation belonging to the dynamics of a large ensemble of such elementary systems.

The centrality of the dynamical rôle of a CD allows us to give a well defined meaning to the sum $\sum_{|\vec{k}|=\omega_0}$ appearing in Eqs. (3.4a) and (3.4b). Quantizing the e.m. field in such domain, that for simplicity we take as a box of size L^3 with $L = \frac{2\pi}{\omega_0}$, the number of independent modes with $|\vec{k}| = \omega_0$ is given by [Ex. $\langle 3.1\rangle$]

$$n(|\vec{k}| = \omega_0) \simeq 4\pi, \qquad (3.5)$$

implying the existence of 4π "independent momentum directions". As a result the meaning of the sum $\sum_{|\vec{k}|=\omega_0}$, that we are seeking, is simply

$$\sum_{|\vec{k}|=\omega_0} = \int d\Omega_{\vec{k}}, \tag{3.6}$$

an integral over the solid angle of the e.m. field wave-numbers $\vec{k}$.

3.2 The dynamics of a single Coherence Domain

According to the above considerations, if we restrict our attention to a single CD inside it we can neglect to a first approximation any space-dependence, and the CE's (3.4) get further simplified. In fact setting $\chi_i(\vec{x}, t) \simeq \frac{1}{\sqrt{V_{CD}}}\chi_i(t)$ we have

$$i\dot{\chi}_2(t) = eJ\sqrt{\frac{N}{V}}\frac{1}{\sqrt{2\omega_0}}\sum_r \int d\Omega_{\vec{k}}\epsilon^*_{\vec{k}r,1}\alpha^*_{\vec{k}r}\chi_1(t), \tag{3.7a}$$

$$i\dot{\chi}_1(t) = eJ\sqrt{\frac{N}{V}}\frac{1}{\sqrt{2\omega_0}}\sum_r \int d\Omega_{\vec{k}}\epsilon_{\vec{k}r,1}\alpha_{\vec{k}r}\chi_2(t), \tag{3.7b}$$

$$-\frac{1}{2\omega_0}\ddot{\alpha}_{\vec{k}r} + i\dot{\alpha}_{\vec{k}r} - \frac{e^2}{\omega_0}\left(\frac{N}{V}\right)\lambda\alpha_{\vec{k}r} = eJ\sqrt{\frac{N}{V}}\frac{1}{\sqrt{2\omega_0}}\epsilon^*_{\vec{k}r,1}\chi^*_2(t)\chi_1(t), \tag{3.7c}$$

with the normalization condition

$$|\chi_1|^2 + |\chi_2|^2 = 1. \tag{3.8}$$

The system (3.7) can be brought to the following standard form:

$$\dot{\chi}_2 = -igA^*\chi_1 \tag{3.9a}$$

$$\dot{\chi}_1 = -igA\chi_2 \tag{3.9b}$$

$$\frac{i}{2}\ddot{A} + \dot{A} + i\mu A = -ig\chi^*_2\chi_1 \tag{3.9c}$$

where we have introduced the dimensionless time $\tau = \omega_0 t$, and have defined:

$$A(\tau) = \sum_r \left(\frac{3}{8\pi}\right)^{\frac{1}{2}} \int d\Omega_{\vec{k}}\alpha_{\vec{k}r}\epsilon_{\vec{k}r,1} \tag{3.10}$$

$$g = eJ\left(\frac{8\pi}{3}\right)^{\frac{1}{2}}\left(\frac{N}{2V\omega_0^3}\right)^{\frac{1}{2}} \tag{3.11}$$

$$\mu = \frac{e^2 \lambda}{\omega_0{}^2} \left(\frac{N}{V} \right).$$ (3.12)

Note that both g and μ are dimensionless quantities.

One can show without any difficulties that the differential system (3.9) possesses the following conserved quantities [Ex. $\langle 3.2 \rangle$]:

$$1 = \chi_1^* \chi_1 + \chi_2^* \chi_2,$$ (3.13)

$$Q = A^* A + \frac{i}{2}(A^* \dot{A} - \dot{A}^* A) + \chi_1^* \chi_1,$$ (3.14)

$$H = Q + \frac{1}{2} \dot{A}^* \dot{A} + \mu A^* A + g(A^* \chi_2^* \chi_1 + A \chi_1^* \chi_2).$$ (3.15)

The quantity Q, quadratic in the field amplitudes, will be for convenience called the "momentum of the system", while in the quantity H [eq. (3.15)] one can without much ado recognize the Hamiltonian (divided by N_{CD}) of the classical system.

The first question we wish to ask the differential system (3.9) is under what conditions starting with an initial configuration (N_{CD} is the number of particles in a coherence domain)

$$A(0) \sim \frac{1}{\sqrt{N_{CD}}}, \qquad \chi_1(0) \sim \frac{1}{\sqrt{N_{CD}}}, \qquad \chi_2(0) \sim 1,$$ (3.16)

corresponding to the perturbative QED ground state (PGS), the system will "run away" from it, toward some new configuration where both A and χ_1 will be appreciably different from zero. It is clear that the answer we seek is contained in the short-time behaviour of the equation

$$\frac{i}{2} \dddot{A} + \ddot{A} + i\mu\dot{A} + g^2 A = 0,$$ (3.17)

that can be obtained by differentiating (3.9c) and substituting in it the r.h.s. of (3.9b). To the linear eq. (3.17) one associates the algebraic equation

$$\frac{p^3}{2} - p^2 - \mu p + g^2 = 0,$$ (3.18)

which has in general three complex roots. A "run away" appears for those values of μ and g^2 for which (3.18) has only one real root, the other two being complex conjugates. It is precisely due to the existence of such complex conjugate roots that in the short-time behaviour of $A(\tau)$ an exponentially increasing term appears that overcomes its very small $O(\frac{1}{\sqrt{N_{CD}}})$ initial value. A simple analysis shows that a run-away exists when, for μ fixed,

$$g^2 > g_c{}^2, \tag{3.19a}$$

with

$$g_c{}^2 = \frac{8}{27} + \frac{2}{3}\mu + \left[\frac{4}{9} + \frac{2}{3}\mu^2\right]^{\frac{3}{2}}. \tag{3.19b}$$

When (3.19a) is verified, the differential equations (3.9) predict that the system will evolve into a state very different from the incoherent Perturbative Ground State (PGS), where the "oscillators" associated with the wave-field χ_1 and the e.m. radiation field A perform zero-point incoherent oscillations, the assumption being usually made that they are only "weakly", perturbatively coupled. This new physical situation, which we shall explore in a moment, will be denoted "strong coherence".

What will happen, instead, if (3.19a) fails to be satisfied? Will the incoherent perturbative ground state remain a good approximation? We shall see that when N_{CD} is not too large a number, and this happens for very large e.m. fields frequencies ω_0, it will prove energetically favourable for the system to access a state where the matter and the e.m. field acquire small but not completely negligible $[O(\frac{1}{\sqrt{N_{CD}}})]$ coherent amplitudes, a situation which we shall refer to as "weak coherence".

Among the solutions of the differential system (3.9) a special status is enjoyed by those "paths" that have the form (a_i, $\mathcal{A}$ real and positive)

$$\begin{aligned} \chi_i &= a_i e^{i\theta_i(\tau)}, \\ A &= \mathcal{A} e^{i\phi(\tau)}, \end{aligned} \tag{3.20}$$

whose time-dependence is only carried by their phase-factors.

Substituting (3.20) in (3.9), we have

$$\dot{\theta}_2 = -g\mathcal{A}\frac{a_1}{a_2}e^{i\psi(\tau)} \tag{3.21a}$$

$$\dot{\theta}_1 = -g\mathcal{A}\frac{a_2}{a_1}e^{-i\psi(\tau)} \tag{3.21b}$$

$$-\frac{\dot{\phi}^2}{2} + \dot{\phi} + \mu = -g\frac{a_1 a_2}{\mathcal{A}}e^{-i\psi(\tau)}, \tag{3.21c}$$

where

$$\psi(\tau) = \theta_1(\tau) - \theta_2(\tau) - \phi(\tau). \tag{3.22}$$

The conditions (3.13) and (3.14) can be expressed by setting $a_2 = \cos\alpha$ and $a_1 = \sin\alpha$ $(0 \le \alpha \le \frac{\pi}{2})$, and

$$\mathcal{A}^2[1 - \dot\phi] + \sin^2\alpha = Q_0. \qquad (3.23)$$

In general we shall fix the value of Q_0 at its "perturbative" initial value, which is of order $\frac{1}{N_{CD}}$.

We easily check that no consistent solution of (3.21) exists unless

$$\psi(\tau) = 0 \text{ or } \pi; \qquad (3.24)$$

when this happens the system becomes

$$\dot\theta_2 = -g\epsilon\mathcal{A}\tan\alpha, \qquad (3.25a)$$

$$\dot\theta_1 = -g\epsilon\mathcal{A}\cot\alpha, \qquad (3.25b)$$

$$\dot\phi = 1 \pm \sqrt{1 + 2\mu + g\epsilon\frac{\sin(2\alpha)}{\mathcal{A}}}, \qquad (3.25c)$$

where $\epsilon = \pm 1$, according to whether $\psi = 0$ or π.

Another consequence of (3.24) is

$$\dot\phi = \dot\theta_1 - \dot\theta_2. \qquad (3.26)$$

Noting that the five unknown quantities α, $\dot\theta_1$, $\dot\theta_2$, $\mathcal{A}$, $\dot\phi$ obey five equations [(3.23), (3.25) and (3.26)] we conclude that the "stationary path(s)" is completely determined.

It will now be demonstrated that the solution with $\epsilon = \pm 1$ are those with maximum and minimum energy respectively. Let us consider the functional

$$\mathcal{H} = H + \lambda(Q - Q_0) + \rho(\chi_1^*\chi_1 + \chi_2^*\chi_2 - 1)$$

$$= Q(1 + \lambda) + \rho(\chi_1^*\chi_1 + \chi_2^*\chi_2 - 1) + \frac{1}{2}\dot{A}^*\dot{A} + \mu A^*A + g(\chi_2^*\chi_1 A^* + \chi_1^*\chi_2 A) - \lambda Q_0.$$

$$(3.27)$$

The stationarity equations for $\mathcal{H}$ are

$$\frac{\delta\mathcal{H}}{\delta\chi_1^*} = (1 + \lambda)\chi_1 + \rho\chi_1 + gA\chi_2 = 0, \qquad (3.28a)$$

$$\frac{\delta\mathcal{H}}{\delta\chi_2^*} = \rho\chi_2 + gA^*\chi_1 = 0, \qquad (3.28b)$$

$$\frac{\delta\mathcal{H}}{\delta A^*} = (1 + \lambda)[A + i\dot{A}] - \frac{\ddot{A}}{2} + \mu A + g\chi_2^*\chi_1 = 0. \qquad (3.28c)$$

Exploiting the equations of motion (3.9), the stationarity equations (3.28) become:

$$i\dot{\chi}_1 + (1 + \lambda + \rho)\chi_1 = 0, \qquad (3.29a)$$

$$i\dot{\chi}_2 + \rho\chi_2 = 0, \qquad (3.29b)$$

$$-\ddot{A} + (2 + \lambda)i\dot{A} + (1 + \lambda)A = 0, \qquad (3.29c)$$

demonstrating that in order to make the Hamiltonian stationary χ_1, χ_2 and A must indeed have the form (3.20) [Ex. $\langle 3.3 \rangle$]. Computing the Hamiltonian (3.15) for the solutions (3.25) we get:

$$H = Q_0 + \mathcal{A}^2 \left(\frac{\dot{\phi}^2}{2} + \mu \right) + \epsilon g \mathcal{A} \sin(2\alpha), \qquad (3.30)$$

showing that $\epsilon = -1$ defines the state of minimum energy, while in the state defined by $\epsilon = +1$ the energy is maximum.

This conclusion is rather interesting for it shows that, irrespective of whether "strong coherence" sets in (and for this, we know, (3.19) must be satisfied), the incoherent PGS whose energy is just Q_0, will never be the real ground state — the state that minimizes the energy. The real ground state is instead coherent, and corresponds to the stationary path with $\psi = \theta_1 - \theta_2 - \phi = \pi$. As a consequence we learn that in general the order that sets in at zero temperature for matter systems coupled to the radiation e.m. field is always coherent irrespective of the strength of their coupling.

It serves no useful purpose to analyse in detail here the solutions of the algebraic system (3.23), (3.25) and (3.26) that fully characterizes the "stationary path" of the CE's for the two-level system; such analysis will be carried out, with much more profit, in a number of concrete, and generally applicable situations in the next Chapters.

3.3   Correcting some fallacies from Laser physics

We must now carry out a long due discussion on the relationship between what we have done so far and the great developments of Laser Physics (LP). This discussion is all the more necessary due to the general ban of the ideas pursued in this book by the eminent U.S. physicist Philip W. Anderson, based precisely on results from LP[1].

What in fact we are trying to investigate in this book is the possible **spontaneous** emergence of non-trivial coherent e.m. fields in the systems of condensed matter, similar to what happens in the cavities of Maser and Lasers, when "pumped" in particular and careful ways.

According to Anderson it is just the **contrived** conditions that are needed to produce the coherent e.m. fields of Maser and Lasers, that expose the delusion of those who conceive of the possibility that such behaviours may emerge **spontaneously** in particular condensed matter systems. In the usual theoretical analysis of LP one deals with differential systems — the Bloch equations — that have close analogies with our CE's but for one important difference: in LP an approximation is universally made — the slowly-varying envelope approximation — according to which the frequency spectrum of the "envelope amplitudes" $\alpha_{\vec{k}r}(t)$ $(\alpha_{\vec{k}r}^*(t))$ of the e.m. field is concentrated at values $|\omega| \ll \omega_{\vec{k}} = |\vec{k}|$ (recall, in fact, that no such restrictions appear in our analysis). This approximation amounts to neglect in our Eqs. (3.17) and (3.18) the third order term. The quadratic equation resulting from (3.18) is then easily seen to have the solutions

$$p_{1,2} = -\frac{\mu}{2} \pm \sqrt{\frac{\mu^2}{4} + g^2}, \qquad (3.31)$$

which for all values of μ and g are real, thus banning the existence of any run-away. It is for this reason that in LP the only way to obtain a non-trivial classical e.m. field is to start from a different initial state, where the upper level wave-field is "pumped" into the configuration $|\chi_1|^2 = \sin^2 \theta$ with $\theta \neq 0$. It is immediate to check that in the latter case g^2 in (3.17) and (3.18) gets changed into $g^2 \Delta = g^2 \cos(2\theta)$, Δ being the population inversion factor $\frac{N_2-N_1}{N_2+N_1}$.

Substituting $g^2 \Delta$ for g^2 in (3.31) we find that the system "runs away" for $-g^2 \Delta > 0$ (μ is usually negligible in LP), i.e. when the population of the upper level 1 exceeds that of the lower level 2, **irrespective of the value of g**, and — from (3.11) — of the value of density, coupling and frequency. Thus we may well call the coherence achieved in LP "brute force coherence".

On the other hand, if one does not make the slowly varying approximation, for $\mu \simeq 0$ and in the uninverted situation ($\Delta = 1$), the system "runs away" if [see eq. (3.19)]

$$g^2 \geq g_c^2 = \frac{16}{27}. \qquad (3.32)$$

This discussion clearly disposes of the fallacy, that we have called "pump fallacy"[2], according to which in Condensed Matter Physics (CMP) no coherent e.m. field can be generated for the obvious absence of a pumping mechanism that preliminarily populates the excited level in excess of the ground state level. In CMP, on the other hand, this obstacle is subtly removed by appropriate values of density $\left(\frac{N}{V}\right)$, coupling J, and frequency ω_0 that allow (3.32) to be satisfied.

But there is another important fallacy that LP would seem to throw against the spontaneous emergence of coherent e.m. fields: we may call it the "cavity fallacy".

In LP, in fact, in order for the coherent e.m. field to develop the atomic systems must be contained in a well tuned optical cavity, designed so as to avoid that the photons generated in the atomic transitions leave the system, and the process be interrupted.

No obvious cavity — Anderson argues — exists in condensed matter systems, so that the coherent e.m. field may be spontaneously confined within condensed matter. Again, these arguments are based on the "slowly varying envelope approximation", according to which the frequencies ω of the photons in the space domain occupied by the radiating atoms are very close to their vacuum frequency $\omega_0 = \omega_{\vec{k}} = |\vec{k}|$. Without making this approximation such frequency is given instead by:

$$\omega = \omega_0(1 - \dot{\phi}) = \omega_0\sqrt{1 - g\frac{\sin(2\alpha)}{\mathcal{A}}} < \omega_0, \tag{3.33}$$

where we have used (3.25c) for the CGS ($\epsilon = -1$), and have set $\mu = 0$. Thus in the CGS the strong coupling between matter and the e.m. field shifts the e.m. field frequency downwards, causing the interface between matter and vacuum to act as a total reflection mirror for the e.m. field, that spontaneously traps the e.m. field inside matter. No finely contrived cavity is thus needed !

We may therefore conclude that the arguments, that have been generally considered very strong, to exclude e.m. coherent fields from any fundamental analysis of condensed matter systems turn out to be simply wrong, being based on the "slowly varying envelope approximation" that, while adequate for most systems considered in LP, is seen to be in general inapplicable to condensed matter systems.

3.4   On the photon "mass-term"

In the preceding analysis we have seen that an important rôle in the CE's is played by the Hamiltonian term $H_{rad}^{(2)}$, which has been introduced in Chapter 1 [Eq. (1.73)]

$$H_{rad}^{(2)} = e^2\lambda \int_{\vec{x},\alpha} \vec{A}(\vec{x},t)^2 \Psi^\dagger(\vec{x},\alpha,t)\Psi(\vec{x},\alpha,t). \tag{3.34}$$

In Chapter 2 [Eq. (2.52)] we have observed that on account of the slow space variation of the density operator $\int_\alpha \Psi^\dagger(\vec{x},\alpha,t)\Psi(\vec{x},\alpha,t)$, $H_{rad}^{(2)}$ gives rise to an effective photon "mass-term" $\int_{\vec{x}} \mu_\gamma^2 \vec{A}(\vec{x},t)^2$, the mass-squared being given by

$$\mu_\gamma^2 = e^2\left(\frac{N}{V}\right)\lambda, \tag{3.35}$$

which in general depends on the frequency.

It is now our purpose to get some better idea regarding the size and the sign of λ, which in spite of the form of (3.35) is not necessarily positive. In Section (1.3) we have argued that λ receives two main contributions, one of which from the "minimal shift" in the free Hamiltonian H_0

$$\vec{p}_i \rightarrow \vec{p}_i + e_i \vec{A}(\vec{x}_i, t) \tag{3.36}$$

of the momenta of the charges e_i comprising the individual atomic (or molecular) system. For definiteness' sake let us consider one such system composed of Z electrons moving around a massive nucleus of opposite charge, which on account of its large mass interacts only negligibly with the e.m. field.

The single-particle Hamiltonian will thus be written (m_e is the electron mass):

$$
\begin{aligned}
H &= \sum_{r=1}^{Z} \frac{\left(\vec{p}_r - e\vec{A}(\vec{x}_r, t) \right)^2}{2m_e} \\
&= H_0 - \frac{e}{m_e} \sum_{r=1}^{Z} \vec{p}_r \cdot \vec{A}(\vec{x}_r, t) + \frac{e^2}{2m_e} \sum_{r=1}^{Z} \vec{A}(\vec{x}_r, t) \cdot \vec{A}(\vec{x}_r, t).
\end{aligned}
\tag{3.37}
$$

In the "dipole-approximation", that neglects the space-variations of the e.m. field over atomic distances (as is quite reasonable considering that the e.m. wave-lengths involved are much larger than the atomic dimensions), we can rewrite (3.37) as

$$H = H_0 + e\vec{J} \cdot \vec{A}(\vec{x}, t) + \frac{e^2}{2m_e} Z \vec{A}(\vec{x}, t)^2, \tag{3.38}$$

where the current operator in the internal atomic space is simply given by

$$\vec{J} = -\sum_{r=1}^{Z} \frac{\vec{p}_r}{m_e}. \tag{3.39}$$

From (3.38) we can immediately obtain the value of λ_{SG} coming from the atomic Hamiltonian (well known in scalar QED as the "seagull term") as

$$\lambda_{SG} = \frac{1}{2m_e} Z. \tag{3.40}$$

But, as argued in Chapter 1, this is not the only second order term of the effective e.m. Hamiltonian, for there are additional terms due to virtual, dispersive interactions that arise from the iterations of the first order term appearing in (3.38).

Using second order perturbation theory such terms can be written as:

$$\dot{H}^{(2)}(\vec{x}, t) = -\frac{ie^2}{2} \int_{-\infty}^{+\infty} d\tau \, \langle \alpha | \, T \left[\vec{J}(t + \frac{\tau}{2}) \cdot \vec{A}(\vec{x}, t + \frac{\tau}{2}) \, \vec{J}(t - \frac{\tau}{2}) \cdot \vec{A}(\vec{x}, t - \frac{\tau}{2}) \right] | \alpha \rangle,$$
$$(3.41)$$

where $| \alpha \rangle$ is the atomic state which the atomic systems find themselves in. Note that we envisage here a situation in which the Hamiltonian $H^{(2)}$ may depend on time as a result of a dynamical evolution from the initial "perturbative QED vacuum" (for which $| \alpha \rangle$ is just the ground state $| 0 \rangle$ of the atomic system, and $\vec{A}$ is the vacuum e.m. mode with $\omega_{\vec{k}} = |\vec{k}|$) to a coherent ground state, where $| \alpha \rangle$ is a well defined superposition of the atomic levels and the e.m. field frequency $\omega_{\vec{k}} \neq |\vec{k}|$. Writing now as usual

$$\vec{A}(\vec{x}, t) = \sum_{\vec{k}r} \frac{1}{\sqrt{2\omega_{\vec{k}} V}} \left[\alpha_{\vec{k}r}(t) \vec{\epsilon}_{\vec{k}r} e^{-i(\omega_{\vec{k}} t - \vec{k} \cdot \vec{x})} + \text{c.c.} \right], \tag{3.42}$$

and inserting a complete system $| n \rangle$ of eigenstates of the atomic Hamiltonian H_0, for the frequency dependent coefficient of the term $\sum_r \frac{1}{\omega_{\vec{k}}} \alpha^*_{\vec{k}r} \alpha_{\vec{k}r}$, that according to (2.52) builds up $H_{rad}^{(2)}$, we can obtain without much difficulty

$$\lambda_{\vec{k}} = -\sum_{nr} \frac{(E_n - E_\alpha)}{(E_n - E_\alpha)^2 - \omega_{\vec{k}}^2} \langle \alpha | \, \vec{\epsilon}_{\vec{k}r} \cdot \vec{J} \, | n \rangle \langle n | \, \vec{\epsilon}_{\vec{k}r}^* \cdot \vec{J} \, | \alpha \rangle$$

$$= -\sum_{nr} \frac{(E_n - E_\alpha)}{(E_n - E_\alpha)^2 - \omega_{\vec{k}}^2} \epsilon_{\vec{k}r,i} \epsilon_{\vec{k}r,j}^* \sum_{st} \langle \alpha | \, \frac{p_{si}}{m_e} \, | n \rangle \langle n | \, \frac{p_{kj}}{m_e} \, | \alpha \rangle$$

$$= -\frac{1}{2m_e} \sum_{nrst} \frac{(E_n - E_\alpha)^2}{(E_n - E_\alpha)^2 - \omega_{\vec{k}}^2} \epsilon_{\vec{k}r,i} \epsilon_{\vec{k}r,j}^* \tag{3.43}$$

$$[\langle \alpha | \, p_{si} \, | n \rangle \langle n | \, x_{tj} \, | \alpha \rangle - \langle \alpha | \, x_{si} \, | n \rangle \langle n | \, p_{tj} \, | \alpha \rangle]$$

where in the last line we have used the Heisenberg equation $\frac{\vec{p}}{m_e} = \dot{\vec{x}} = i[H, \vec{x}]$.

A simple algebraic rearrangement, and the use of the canonical commutation relations

$$[p_{si}, x_{tj}] = -i\delta_{st}\delta_{ij} \tag{3.44}$$

allow us to write :

$$\lambda_{\vec{k}} = -\frac{Z}{2m_e} - \sum_{nr} \frac{\omega_{\vec{k}}^2}{(E_n - E_\alpha)^2 - \omega_{\vec{k}}^2} \frac{| \langle \alpha | \, \vec{\epsilon}_{\vec{k}r} \cdot \vec{J} \, | n \rangle |^2}{(E_n - E_\alpha)}. \tag{3.45}$$

The most remarkable aspect of (3.45) is the frequency independent term $-\frac{e^2}{2m_e} Z$ that exactly cancels λ_{SG}. This result is physically quite transparent and eminently reasonable, for it shows that the e.m. field does not interact with all electrons of the atom equally, as the sea-gull terms would have it. It is indeed obvious that the deeply bound electrons of high Z-atoms are totally segregated from e.m. modes of comparatively low frequency. The total second order interaction term, that we can now write

$$\lambda = -e^2 \sum_{nr} \frac{\omega_{\vec{k}}^2 (E_n - E_\alpha)}{(E_n - E_\alpha)^2 - \omega_{\vec{k}}^2} \sum_{ij} [\langle \alpha | \vec{\epsilon}_{\vec{k}r} \cdot \vec{x}_j | n \rangle \langle n | \vec{\epsilon}_{\vec{k}r} \cdot \vec{x}_i | \alpha \rangle] \qquad (3.46)$$

shows in fact very clearly that transitions with energy differences $(E_n - E_\alpha) \gg \omega_{\vec{k}}$ are, as expected, highly suppressed.

It is worth recalling here that an interesting result, derived at the beginning of the seventies by K. Hepp and E. H. Lieb[3], on the possible "superradiant" phase-transition of systems of N two-level atoms, was challenged in the mid seventies in a series of papers[4] that, using for the second order Hamiltonian $H_{rad}^{(2)}$ the sea-gull term only (neglected in the analysis of Hepp and Lieb), proved such transition to be impossible. Fifteen years later we see that another result, that went in the direction of the ideas being developed in this book, was prematurely discarded, based just on trivially fallacious arguments[5].

From the explicit expression (3.46) we can see that for two-level systems λ will in general be rather small due to the fact that the transition strengths $|\langle n | \vec{J} \cdot \vec{\epsilon} | \alpha \rangle|^2$ are highest between the ground state $|\alpha\rangle$ and the particular excited state comprising the two-level system, and that it is just this transition that should be excluded from the sum appearing in (3.46), in order to avoid double counting. But there are exceptions, as we shall see later.

3.5 The space-structure of the "stationary paths"

We have argued in Section (3.1) that the space-structure of the "stationary paths", and in particular of the CGS ($\epsilon = -1$), can be visualized as an array of Coherence Domains (CD's) separated by interfaces where the e.m. coherent field has small or vanishing amplitudes and likewise the matter field resumes a configuration typical of the perturbative QED ground state. In this Section we are going to give a more precise substance to these expectations.

We start from the CE's of the two-level system in the form (3.3), where we get rid of the laplacian term being of order $\frac{\omega_0^2}{2m_e}$ which is usually much smaller than ω_0, the typical strength of the e.m. interaction term. We rewrite the differential system in the form :

$$i\frac{\partial}{\partial t}\chi_2(\vec{x},\tau) = g\left(\frac{3}{8\pi}\right)^{\frac{1}{2}} \sum_{|\vec{k}|=\omega_0,r} \epsilon^*_{\vec{k}r,1}\alpha^*_{\vec{k}r}(\tau)e^{-i\vec{k}\cdot\vec{x}}\chi_1(\vec{x},\tau), \qquad (3.47a)$$

$$i\frac{\partial}{\partial t}\chi_1(\vec{x},\tau) = g\left(\frac{3}{8\pi}\right)^{\frac{1}{2}} \sum_{|\vec{k}|=\omega_0,r} \epsilon_{\vec{k}r,1}\alpha_{\vec{k}r}(\tau)e^{-i\vec{k}\cdot\vec{x}}\chi_2(\vec{x},\tau), \qquad (3.47b)$$

$$-\frac{1}{2}\ddot{\alpha}_{\vec{k}r} + i\dot{\alpha}_{\vec{k}r} - \mu\alpha_{\vec{k}r} = g\left(\frac{3}{8\pi}\right)^{\frac{1}{2}} \epsilon^*_{\vec{k}r,1} \int_{\vec{x}} e^{-i\vec{k}\cdot\vec{x}}\chi_2^*(\vec{x},\tau)\chi_1(\vec{x},\tau), \qquad (3.47c)$$

which can be rewritten:

$$i\frac{\partial}{\partial t}\chi_2(\vec{x},\tau) = gA^*(\vec{x},\tau)\chi_1(\vec{x},\tau) \qquad (3.48a)$$

$$i\frac{\partial}{\partial t}\chi_1(\vec{x},\tau) = gA(\vec{x},\tau)\chi_2(\vec{x},\tau) \qquad (3.48b)$$

$$-\frac{1}{2}\ddot{A}(\vec{x},\tau) + i\dot{A}(\vec{x},\tau) - \mu A = g\int_{\vec{x}'} G(\vec{x}-\vec{x}')\chi_2^*(\vec{x}',\tau)\chi_1(\vec{x}',\tau)(\vec{x},\tau), \qquad (3.48c)$$

where

$$A(\vec{x},\tau) = \left(\frac{3}{8\pi}\right)^{\frac{1}{2}} \sum_{|\vec{k}|=\omega_0,r} e^{i\vec{k}\cdot\vec{x}}\epsilon_{\vec{k}r,1}\alpha_{\vec{k}r}(\tau), \qquad (3.49)$$

and

$$G(\vec{x}) = \left(\frac{3}{8\pi}\right) \sum_{|\vec{k}|=\omega_0,r} e^{i\vec{k}\cdot\vec{x}}\epsilon_{\vec{k}r,1}\epsilon^*_{\vec{k}r,1}. \qquad (3.50)$$

The integral appearing in the r.h.s. of (3.48c) can be rewritten:

$$\int_{\vec{\xi}} G(\vec{\xi})\chi_2^*(\vec{x}-\vec{\xi},\tau)\chi_1(\vec{x}-\vec{\xi},\tau) \simeq \int_{\vec{\xi}} G(\vec{\xi})\chi_2^*(\vec{x},\tau)\chi_1(\vec{x},\tau) = \chi_2^*(\vec{x},\tau)\chi_1(\vec{x},\tau), \quad (3.51)$$

where use has been made of the conjecture, to be checked "a posteriori", that the terms arising from the Taylor expansion in $\vec{\xi}$ of $\chi_2^*(\vec{x}-\vec{\xi},\tau)\chi_1(\vec{x}-\vec{\xi},\tau)$ can to a good approximation be neglected.

Substituting (3.51) in (3.48c), we find that we have obtained a system of the type (3.9) for every space-dependent amplitude $\chi_i(\vec{x},t)$ and $A(\vec{x},t)$.

However an important point to note is that the constraints (3.13) and (3.14) are now in integral form, and they read

$$\frac{1}{V_{CD}} \int_{\vec{x}} (\chi_1^*\chi_1 + \chi_2^*\chi_2) = 1 \qquad (3.52)$$

and

$$\frac{1}{V_{CD}} \int_{\vec{x}} \left[A^*A + \frac{i}{2}(A^*\dot{A} - \dot{A}^*A) + \chi_1^*\chi_1 \right] = Q_0. \tag{3.53}$$

We can now solve for the stationary path of minimum energy

$$\chi_i(\vec{x}, \tau) = a_i(\vec{x})e^{i\theta_i(\vec{x}, \tau)} \tag{3.54}$$

$$A(\vec{x}, \tau) = \mathcal{A}(\vec{x})e^{i\phi(\vec{x}, \tau)}, \tag{3.55}$$

obtaining as usual [see eq. (3.25)]

$$\dot{\phi}(\vec{x}, \tau) = \dot{\theta}_1(\vec{x}, \tau) - \dot{\theta}_2(\vec{x}, \tau) = 2g\mathcal{A}(\vec{x})\cot[2\alpha(\vec{x})] \tag{3.56}$$

and

$$\dot{\phi}(\vec{x}, \tau) = 1 \pm \sqrt{1 + 2\mu - g\frac{\sin[2\alpha(\vec{x})]}{\mathcal{A}(\vec{x})}}, \tag{3.57}$$

which establish between $\mathcal{A}(\vec{x})$ and $\alpha(\vec{x})$ the same kind of relationship that exists in the $\vec{x}$-independent case. As for $\mathcal{A}(\vec{x})$ we can determine its space-structure very easily by noting that from (3.47c) $\alpha_{\vec{k}r} \approx \epsilon^*_{\vec{k}r,1}$. Substituting in (3.49) we immediately conclude that

$$A(\vec{x}, \tau) = A(\tau)\left|\frac{G(\vec{x})}{G(\vec{0})}\right|, \tag{3.58}$$

where $G(\vec{x})$ has been defined in (3.50) and $A(\tau)$ essentially coincides with the space-independent solution.

In order to determine the space-modulation of the e.m. field amplitude $A(\vec{x}, \tau)$ over the whole CGS, we proceed as follows: we start from a single CD, and build up the complete CGS by "nucleating" the appropriate number of CD's. Note that this construction follows very closely the actual physical formation of the condensed system from the aggregation (nucleation) of a myriad of elementary units, which are nothing but the CD's of the relevant QED coherent process.

For a single CD of radius r_0 the definition (3.50) leads us to the simple form $\frac{G(\vec{x})}{G(\vec{0})} = \frac{\sin\omega_0 r}{\omega_0 r}$ (r is the radius of a spherical CD), thus we may rewrite (3.58)

$$A(\vec{x}, \tau) = A(0)\frac{\sin\omega_0 r}{\omega_0 r}e^{-i\omega_r t} \qquad (r < r_0) \tag{3.59}$$

$[\omega_r = \omega_0(1 - \dot{\phi})]$. Outside the CD $(r > r_0)$ $\vec{A}(\vec{x}, \tau)$ obeys the free-field equation

$$\Box \vec{A}(\vec{x}, \tau) = 0, \tag{3.60a}$$

which, factoring out the time-dependence of (3.59), can be written

$$\left(-\omega_r^2 - \vec{\nabla}^2\right) \vec{A}(\vec{x}) = 0 \qquad (r > r_0). \tag{3.60b}$$

We now notice that at the border of the CD $(r = r_0)$ the angular part of the laplacian $-\vec{\nabla}^2$, $\frac{1}{r_0^2}\vec{L}^2$, when applied to $\vec{A}(\vec{x}, \tau)$, yields approximately $|\vec{k}|^2 \vec{A}(\vec{x}, \tau)|_{|\vec{x}|=r_0}$ ($|\vec{k}| = \omega_0$), a result stemming from the fact that, at the interface with the perturbative vacuum, the e.m. modes have their wave numbers $\vec{k}$ tangential to the interface. Using this fact and the spherical symmetry of the problem, eqs. (3.60) become:

$$\frac{d^2}{dr^2}(rA) - (\omega_0^2 - \omega_r^2)(rA) = 0 \qquad (r > r_0). \tag{3.61}$$

In order to proceed any further we must know the precise value of r_0 . Little thought is required to realize that r_0 is determined by joining (3.59) together with its first radial derivative to the exponentially decaying solution of (3.61). A simple calculation for $\omega_r \ll \omega_0$, as is usually the case, yields

$$r_0 \approx \frac{3\pi}{4\omega_0}, \tag{3.62}$$

and the "evanescent e.m. wave-field" is given by

$$A(r) \approx \frac{A(0)}{\sqrt{2}} \frac{e^{-\sqrt{\omega_0^2 - \omega_r^2}(r-r_0)}}{\omega_0 r}. \tag{3.63}$$

The radial profile is depicted in Fig. 3.1. We remark that these simple calculations shed light on the physical mechanism that keeps the e.m. radiation "trapped" inside the region where the coherent dynamical evolution with the matter field occurs.

As discussed in Section (3.3), it is just the large frequency shift (from ω_0 to ω_r) that the e.m. field experiences inside the CD that prevents it from "leaking out" into the vacuum, where for a propagating field both the frequency and the wave-number should equal ω_0. Eq. (3.63) is the simple mathematical expression that exposes the "cavity fallacy" discussed in Section (3.3).

We are now ready to determine the spatial profile of the "stationary path" where many CD's have been brought to be closely packed, at a minimum interdomain distance (between the CD centers) equal to $2r_0$. In this way within each domain one

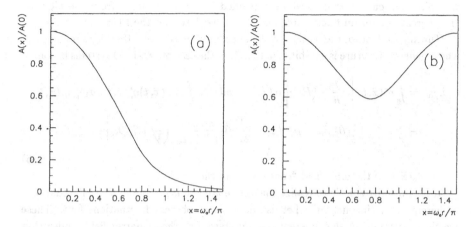

Fig. 3.1. The radial profile of the e.m. amplitude (a) in the isolated CD, (b) in the bulk, as a function of the variable $x = \frac{\omega_0 r}{\pi}$.

has a superposition of the "inner" field and of the evanescent tail of the neighbouring domains. To a satisfactory approximation one has thus within each given domain $(\omega_0 r = \pi x, \; 0 < x < \frac{3}{4})$

$$A(x) = A(0)F(x) = A(0)\left[\frac{\sin \pi x}{\pi x} + \frac{\sqrt{2}}{\pi}\frac{e^{-\pi(\frac{3}{4}-x)}}{3-2x}\right]. \tag{3.64}$$

Note that the close packing of the domains proves energetically advantageous since, for instance, at $x = \frac{3}{4}$ in such configurations an atomic system gains an energy four times larger than in the isolated domain.

3.6   The quantum fluctuations

The energy of the coherent ground state, which inside the CD corresponds to the "stationary path" defined in the eqs. (3.23), (3.25) and (3.26) with $\epsilon = -1$, according to (3.30) is lower than the energy E_0 of the perturbative ground state $N\omega_0 Q_0$ by the amount

$$\Delta E = E_0 - E_{CGS} = N\omega_0\left[\mathcal{A}^2\left(\frac{\dot{\phi}^2}{2}+\mu\right) - g\mathcal{A}\sin(2\alpha)\right] = -N\Delta. \tag{3.65}$$

In the space-dependent solutions discussed in the preceding Section the gap Δ is a space-dependent quantity that will be denoted $\Delta(\vec{x})$, that can be interpreted as the minimum energy that must be expended in order to create from the CGS an "incoherent" quantum fluctuation, one that "lives" around the PGS.

Turning, in fact, our attention to the quantum fluctuations, the Lagrangian L_{fluc}, whose general structure is exhibited in (2.57), in the case of two-level systems becomes:

$$L_{fluc} = \int_{\vec{x}} \eta_i^*(\vec{x}, t) \left[i\frac{\partial}{\partial t} - (H_0)_i \right] \eta_i(\vec{x}, t) - eJ\sqrt{N} \int_{\vec{x}} A_{01}(\vec{x}, t) \eta_i^*(\vec{x}, t)\, (\sigma_1)_{ij}\, \eta_j(\vec{x}, t)$$

$$+ \sum_{|\vec{k}|=\omega_0} \sum_r \left[\frac{i}{2}(\beta_{\vec{k}r}^* \dot{\beta}_{\vec{k}r} - \dot{\beta}_{\vec{k}r}^* \beta_{\vec{k}r}) + \frac{1}{2\omega_0}\dot{\beta}_{\vec{k}r}^* \dot{\beta}_{\vec{k}r} - \frac{e^2}{\omega_0}\left(\frac{N}{V}\right)\beta_{\vec{k}r}^* \beta_{\vec{k}r} \right],$$

$$(3.66)$$

where $\vec{A}_0(\vec{x}, t)$ is the e.m. field "condensed" in the CGS.

It is now quite simple to realize that there are two types of quantum fluctuations, "coherent" and "incoherent". Let us analyse the coherent fluctuations first. These are the excitations of the coherent ground state, i.e. those matter field-modes that do not "leave" the coherent ground state and keep a well defined phase relationship with the e.m. condensate. The matter field associated to such fluctuations can be written (note that $\sum_i \chi_i^* \chi_i = 1$):

$$\eta_{i\,coh}(\vec{x}, t) = \sum_{\vec{k}} \frac{a_{\vec{k}}(t)}{\sqrt{V}} e^{i\vec{k}\cdot\vec{x}} \chi_i(\vec{x}, t), \qquad (3.67)$$

which when inserted in (3.66) leads to the free Lagrangian:

$$L_{coh} = \sum_{\vec{k}} a_{\vec{k}}^* \left[i\frac{\partial}{\partial t} - \frac{\vec{k}^2}{2m} \right] a_{\vec{k}}, \qquad (3.68)$$

as one can easily derive from the fact that the classical fields χ_i obey the CE's. The spectrum of the coherent plane-wave excitations is thus

$$\epsilon_{\vec{k}} = \frac{\vec{k}^2}{2m}, \qquad (3.69)$$

showing that, as expected, for the "coherent" fluctuations there is no gap. Repeating the discussion carried out in Sect. (1.4) that led to (1.94) we realize at once that the spectrum of coherent fluctuations has a cut-off for $|\vec{k}| = |\vec{k}_0| \approx \left(\frac{N}{V}\right)^{\frac{1}{3}}$, for the regions resolved at such momenta contain very few elementary systems, and the phase-particle number uncertainty relation (1.14) forbids the matter wave-field to have the necessary well-defined phase, implied by (3.67).

Turning now our attention to the "incoherent fluctuations", they correspond to the fluctuations around the perturbative ground state, for which the e.m. condensate $\vec{A}(\vec{x}, t)$ is either zero or it averages to zero.

Setting

$$\eta_{i\,inc}(\vec{x}, t) = \frac{1}{\sqrt{V}} \sum_{\vec{k}} b_{i\vec{k}}(t) e^{i\vec{k}\cdot\vec{x}} \tag{3.70}$$

we have for the incoherent Lagrangian

$$L_{inc} = \sum_{\vec{k}i} b_{i\vec{k}}^{*} \left[i\frac{\partial}{\partial t} - \left(E_i + \frac{\vec{k}^2}{2m} + \Delta \right) \right] b_{i\vec{k}}, \tag{3.71}$$

which shows that the energy spectrum now possesses a gap Δ that, according to the preceding discussion, represent the energy that one must give the quasi-particle in order to excite it from the CGS to the PGS.

In Sect. (2.3) we have argued that the dynamics of both coherent and incoherent fluctuations is generally perturbative; the interaction Hamiltonian comprises besides the residual electromagnetic interaction

$$H_{em} = eJ \int_{\vec{x}} [A_1(\vec{x}, t) - A_{01}(\vec{x}, t)] \psi_i^{\dagger}(\vec{x}, t) (\sigma_1)_{ij} \psi_j(\vec{x}, t) \tag{3.72}$$

the short-range interaction, neglected so far, that can be generally represented as

$$H_{int} = \frac{1}{2} \int_{\vec{x},\vec{y}} \psi_i^{\dagger}(\vec{x}, t) \psi_i(\vec{x}, t) V(\vec{x} - \vec{y}) \psi_j^{\dagger}(\vec{y}, t) \psi_j(\vec{y}, t), \tag{3.73}$$

where $V(\vec{x} - \vec{y})$ is a generic short-range potential, arising from electrostatic interactions. Writing

$$\psi_i = \chi_i + \eta = \chi_i + \eta_{coh} + \eta_{inc}, \tag{3.74}$$

it is clear that H_{int} will contain cubic and quartic interaction terms between the η's, that can be dealt with following the usual prescriptions of PT (Perturbation Theory). The quadratic terms, on the other hand, are dealt à la Bogoliubov, correcting the spectra (3.69) and (3.71) to allow for "phononic" behaviours, as discussed in Sect. (1.4).

Thus, armed with the tools of Bogoliubov trf. and PT, we can carry out a systematic and complete analysis of the important problem of the dynamics of quantum fluctuations.

3.7   Relaxing from the PGS to the CGS

Due to energy conservation the state, that according to the CE's evolves from the PGS, has the same energy $E_0 = Q_0$ of the PGS which is higher than that of the CGS, the real ground state. If the system were a closed "classical" system it would continue to dynamically evolve as an excited field configuration forever, never coming down to the ground state. But for an open QFT system, we know, the situation is completely different due to the quantum fluctuations that allow the excited state to dissipate its energy and finally "relax" to the ground state.

In this Section we wish to "conceptualize" this important relaxation process within our framework. In order to do this let us represent the state that evolves from the PGS at $t = 0$ [$A(0) \simeq \frac{1}{\sqrt{N_{CD}}}$, $\chi_1(0) \simeq \frac{1}{\sqrt{N_{CD}}}$, $\chi_2(0) \simeq 1$] as

$$A(t) = \mathcal{A}e^{i\phi(t)} + \delta A(t), \qquad (3.75a)$$

$$\chi_i(t) = a_i e^{i\theta_i(t)} + \delta\chi_i(t). \qquad (3.75b)$$

where the definitions (3.20) have been used.

From the standard system (3.9) we see that δA and $\delta\chi_i$ obey the differential equations:

$$\delta\dot{\chi}_2 = -ig(\delta A^*\chi_1 + A^*\delta\chi_1), \qquad (3.76a)$$

$$\delta\dot{\chi}_1 = -ig(\delta A\chi_2 + A\delta\chi_2), \qquad (3.76b)$$

$$i\delta\ddot{A} + \delta\dot{A} + i\mu\delta A = -ig(\delta\chi_2^*\chi_1 + \chi_2^*\delta\chi_1), \qquad (3.76c)$$

a non-linear differential system for the "excitation" fields δA and $\delta\chi_i$, that one can in general solve numerically.

However, without carrying out any computation we know already that both δA and $\delta\chi_i$, due to the stability of the CGS, have a spectrum that comprises only real frequencies. An analysis of the linearized system, obtained from (3.76) by keeping only those terms which are first order in δA and $\delta\chi_i$, [Ex. $\langle 3.4 \rangle$], shows that δA possesses three discrete frequencies $p_j\omega_0$ that, beating against the e.m. condensate $\mathcal{A}e^{i\phi\omega_0 t}$ of the CGS, produce amplitude oscillations with the frequencies

$$\Omega_j = \omega_0(p_j - \dot{\phi}), \qquad (3.77)$$

that are well known in Laser Physics and are called the Rabi frequencies. These oscillations represent the typical dynamical behaviours of the excited solutions of the CE's.

How will the excess energy contained in these configurations get dissipated? There are two classes of processes that contribute to the relaxation of the excited state to the CGS. One class contains all **direct** transition processes to the CGS through the emission of e.m. radiation, the other is made up by all perturbative processes in which the deexcitation occurs through the emission of quantum fluctuations, both matter fluctuations and photons, that will heat the system until they are dissipated to the surrounding "heat bath", thus driving the system to the CGS (of course, at the temperature of the bath).

In order to understand better the general nature of the relaxation process we consider the matrix element of the residual e.m. interaction Hamiltonian (3.72) between the CGS and the PGS. One has $\left[\sigma_1(t) = \begin{pmatrix} 0 & e^{i\omega_0 t} \\ e^{-i\omega_0 t} & 0 \end{pmatrix} \right]$

$$\begin{aligned} \langle PGS | \, H_{em} \, | CGS \rangle &= eJ \langle PGS | \int_{\vec{x}} \chi^\dagger \sigma_1(t) \chi A_1 \, | CGS \rangle \\ &= eJ \int_{\vec{x}} \chi^\dagger_{PGS} \sigma_1(t) \chi_{CGS} A_1 \langle PGS | CGS \rangle, \end{aligned} \tag{3.78}$$

where χ_{PGS} and χ_{CGS} are the matter fields in the PGS and in the CGS respectively; from (3.75) one has

$$\chi_{PGS} - \chi_{CGS} = \delta\chi. \tag{3.79}$$

Let us now compute the scalar product appearing in (3.78). Recalling that, neglecting the space-dependences, both $| CGS \rangle$ and $| PGS \rangle$ are coherent states of the harmonic oscillators associated with the space independent mode with amplitude $\sqrt{N}\chi_{CGS}$ and $\sqrt{N}\chi_{PGS}$ respectively, taking account of the scalar product (1.54) we obtain

$$\langle PGS | CGS \rangle = \exp\left\{ -\frac{N}{2} \left[|\delta\chi|^2 - (\chi^*_{PGS}\chi_{CGS} - \chi^*_{CGS}\chi_{PGS}) \right] \right\}. \tag{3.80}$$

This simple result is very important for it show that macroscopically ($N \to \infty$) different "classical" configurations are in fact orthogonal and no global transition can occur between them. This means that the only way in which the relaxation processes can take place is by breaking up the system in small domains comprising

$$N_0 \simeq \frac{1}{|\delta\chi|^2} \tag{3.81}$$

elementary systems (in which the scalar product (3.80) is of order one) and allowing them to emit independently electromagnetic radiation and matter fluctuations.

To get some idea of the main features of this type of process let us compute the transition amplitude for a photon of momentum $\vec{k}$ ($|\vec{k}| \neq \omega_0$), energy $\omega_{\vec{k}}$ and polarization $\vec{\epsilon}_{\vec{k}r}$. From (3.78) we have

$$A_{\vec{k},\vec{\epsilon}_{\vec{k}r}} = -i \lim_{T\to\infty} \langle CGS | \int_{-\frac{T}{2}}^{\frac{T}{2}} H_{em} dt \, | PGS \rangle$$

$$= -i \frac{N_0}{(2\omega_{\vec{k}} V)^{\frac{1}{2}} e} eJ \lim_{T\to\infty} \int_{-\frac{T}{2}}^{\frac{T}{2}} dt e^{i\omega_{\vec{k}} t} \chi_{CGS}^*(t) \sigma_1(t) \chi_{PGS}(t) \epsilon_{\vec{k}r,1} \left(\frac{1}{V_0} \int_{V_0} d^3\vec{x} e^{-i\vec{k}\cdot\vec{x}} \right),$$

$$(3.82)$$

from which we learn that the system will emit radiation of frequencies $\omega_{\vec{k}} = \sqrt{\vec{k}^2 + \mu_{\vec{k}}^2}$ (note that inside matter the photon has in general a non-zero mass) that belong to the spectrum of Rabi oscillations. Furthermore the space integral of (3.82) indicates that the $\vec{k}$-spectrum is cut-off at $|\vec{k}|_{max} \simeq \left(\frac{2\pi}{a}\right) \frac{1}{N_0^{\frac{1}{3}}}$.

From (3.82) the rate for the mode $\vec{k}$ is easily computed to be:

$$R_{\vec{k}} = \frac{N_0^2}{\exp(2)} \frac{e^2 J^2}{2\omega_{\vec{k}} V} \frac{2}{3} (2\pi) \sum_j |c_j|^2 \delta(\omega_{\vec{k}} - \Omega_j),$$

$$(3.83)$$

where Ω_j are the frequencies of the Rabi oscillations and c_j their amplitudes. To sum over the $\vec{k}$-modes we use the continuum approximation $\sum_{\vec{k}} \to \frac{V}{(2\pi)^3} \int d^3\vec{k}$ and obtain for the relaxation time $\tau = \frac{1}{R}$ ($\alpha = \frac{e^2}{4\pi} \simeq \frac{1}{137}$):

$$\frac{1}{\tau} = \frac{4}{3} \alpha J^2 \frac{N_0^2}{\exp(2)} \sum_j |c_j|^2 \sqrt{\Omega_j^2 - \mu_j^2}.$$

$$(3.84)$$

Setting $\Omega_j \simeq \omega_0$, $|c_j|^2 \simeq |\delta\chi|^2 = \frac{1}{N_0}$, $J \simeq 1$, $N_0 \simeq 10^2$ we can give the following rough estimate of the relaxation time τ

$$\tau \simeq \frac{2\pi}{\omega_0},$$

$$(3.85)$$

which shows that the PGS relaxes to the CGS in the period of typical oscillations of the coherent states of the system: a very interesting and physically meaningful finding.

Such a short relaxation time, however, casts doubts on our perturbative calculation based on (3.82), where the limit $T \to \infty$ has been taken. If instead of $T \to \infty$, we fix $T \simeq \tau$ the calculation makes much more sense, the only difference is now that the sharp energy lines of (3.83) acquire a width $\Delta\omega$ of the order of ω_0, as demanded by Heisenberg's principle. Another consequence that is both interesting and physically very reasonable.

The qualitative ideas that we have gained in this Section will be very important for later developments.

This discussion completes the general analysis of two-level systems. The simplicity of the atomic structure of such systems has allowed us to deal explicitly with some conceptual problems that in the concrete systems, which will be analysed in the following Chapters, will constantly arise. It should be stated very clearly that the subject matter of the last Sections is conceptually rather tough: the reader is well advised to make sure that he grasps it, before he embarks in his navigation through the rest of this book.

Exercises of Chapter 3

$\langle 3.1 \rangle$: Derive Eq. (3.5).

Hint In a cubic box of size L^3 the quantized wave numbers are $\vec{k} = \frac{2\pi}{L}\vec{n}$. The density of states is:

$$dN = \frac{L^3}{(2\pi)^3}d^3\vec{k} \simeq \frac{L^3}{(2\pi)^3}4\pi k^2 \Delta k$$

$\langle 3.2 \rangle$: Show that (3.13), (3.14) and (3.15) are integrals of the differential system (3.9).

$\langle 3.3 \rangle$: Prove that Eq. (3.29) are equivalent to Eq. (3.20), obeying (3.22).

$\langle 3.4 \rangle$: Carry out the analysis leading to Eq. (3.77), and find the equations of which the p_j's are solutions.

References to Chapter 3

1. P. W. Anderson, *Basic Notions of Condensed Matter Physics* (Benjamin-Cummings, Menlo Park (Ca), 1984).
2. G. Preparata, *Coherence in QCD and QED*, in *Common Problems and Ideas of Modern Physics*, eds. T. Bressani, B. Minetti and A. Zenoni (World Scientific, Singapore, 1992).
3. K. Hepp and E. H. Lieb, *Ann. of Phys.* **76** (1973) 360; *Phys. Rev.* **A8** (1973) 2517.

4. K. Rzazewski,K. Wodkiewics and W. Zakowics, *Phys. Rev. Lett.* **35** (1975) 432.

   I. Bialynicki-Birula and K. Rzazewski, *Phys. Rev.* **A19** (1979) 301.

5. E. Del Giudice, R. Mele and G. Preparata, *Mod. Phys. Lett. B* **7**, (1993) 1851.

Chapter 4

QED COHERENCE IN THE TWO HELIUM ISOTOPES

4.1 Normal liquid ^{4}He as a roton gas

Liquid ^{4}He is an enormously interesting system, that has been studied in an incredible depth and extension. The phase $(P-T)$ diagram of ^{4}He (just for visualization purposes reported in Fig. 4.1) does show the remarkable and unique fact that ^{4}He (as well as its isotope ^{3}He) below $P = P_c$ ($P_c = 25$ atm for ^{4}He and $P_c = 33$ atm for ^{3}He) remains liquid at the lowest temperatures that modern technology allows us to reach. Across the λ-line, which at the saturated vapor pressure (SVP) corresponds to $T_\lambda = 2.15$ K, the system undergoes a phase transition from a normal liquid to a superfluid, physically realizing the mathematical model of a perfect liquid: another unique feature that ^{4}He shares with its isotope ^{3}He, which however becomes superfluid at temperatures one thousand times smaller.

In this Chapter we will be able to show that the QED coherence we have discussed and analysed in the previous Chapters, leads to a natural and simple explanation of these fascinating facts as well as to an adequate understanding of the process of solidification. However, before we embark in such developments, it appears appropriate to devote this first Section to a general sketch of the fundamental properties of the normal ^{4}He liquid that play such a relevant rôle, as first discovered and discussed by Landau, in the description of the superfluid phase transition.

According to the basic principles adopted in this book, and described and argued in the preceding Chapters, the properties of the normal liquid should be all understood within the well known schemes of approximation of Generally Accepted Condensed Matter Physics (GACMP), based on short-range electrostatic forces, such as Van der Waals attraction and "hard-core" repulsion, which at the atomic level are efficiently

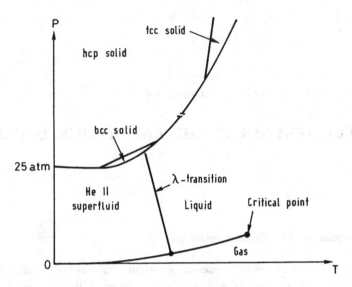

Fig. 4.1. The phase-diagram of ^{4}He

described by two-body Lennard-Jones type of potentials:

$$V_{LJ}(r) = \Delta \left[\left(\frac{r_0}{r} \right)^{12} - 2 \left(\frac{r_0}{r} \right)^{6} \right] .$$ (4.1)

with $\Delta = 10.22$ K and $r_0 = 2.869$Å. Following our discussion in Section (1.3), we associate to liquid ^{4}He the scalar bosonic wave-field $\psi(\vec{x}, t)$ and the Hamiltonian (m_4 is the ^{4}He-mass)

$$H = \int_{\vec{x}} \psi^\dagger(\vec{x}, t) \frac{-\vec{\nabla}^2}{2m_4} \psi(\vec{x}, t) + \frac{1}{2} \int_{\vec{x}\vec{y}} \psi^\dagger(\vec{x}, t) \psi(\vec{x}, t) V(\vec{x} - \vec{y}) \psi^\dagger(\vec{y}, t) \psi(\vec{y}, t),$$ (4.2)

where we would be tempted to substitute for $V(\vec{x} - \vec{y})$, the Lennard-Jones potential (4.1). However, doing this would prevent us from carrying out the Bogoliubov analysis of Section (1.4), for the Fourier transform of V_{LJ} unfortunately does not exist due to its strong singularity for $r = 0$. But this is only a formal difficulty that a proper understanding of the physical situation should be able to do away with. Where does the bug hide? It requires little thought to realize that the difficulty stems from our description of the ^{4}He atom as a pointlike object, as implied by our definition of the local wave-field $\psi(\vec{x}, t)$. Had we properly defined the field $\psi(\vec{x}, \vec{\xi}, t)$, $\vec{x}$ being the center of mass coordinate, $\vec{\xi}$ the internal coordinate so that $\vec{x} + \vec{\xi}$ represents the electronic coordinate, the strong short-distance electronic repulsion (that is responsible for the very singular repulsion implied by the first term of V_{LJ}) would cause the expectation value of the product $\int \vec{\xi}\vec{\eta}\langle \psi^\dagger(\vec{x}, \vec{\xi}, t)\psi(\vec{x}, \vec{\xi}, t)\psi^\dagger(\vec{y}, \vec{\eta}, t)\psi(\vec{y}, \vec{\eta}, t)\rangle$ to strongly vanish for $|\vec{x} - \vec{y}| < 2R_0$, R_0 being the atomic radius. Thus for $V(\vec{x})$ we can take the "pseudopotential" defined by

$$V(r) = \begin{cases} V_{LJ} & \text{for } r > 2R_0 \\ 0 & \text{for } r < 2R_0 , \end{cases} \tag{4.3}$$

avoiding in this way all singularities: the Fourier transform of (4.3) is now perfectly well defined. The Bogoliubov's diagonalization can be carried out exactly as in Section (1.4) with the only difference that the constant λ occurring in that analysis must be replaced by

$$V(\vec{q}) = \int d^3x \, e^{-i\vec{q}\cdot\vec{x}} \, V(r) \simeq 4\pi\Delta(2R_0)^3 \frac{\sin 2qR_0}{2qR_0} \left[\frac{1}{10}\left(\frac{r_0}{2R_0}\right)^{12} - \frac{1}{2}\left(\frac{r_0}{2R_0}\right)^6\right] , \tag{4.4}$$

the Fourier transform of the "pseudopotential"(4.3) [Ex. $\langle 4.1\rangle$]. The transformation equations (1.88) now read

$$E_{\vec{k}} \cosh 2\alpha_{\vec{k}} = \frac{\vec{k}^2}{2m_4} + \left(\frac{N}{V}\right) V(\vec{k})$$

$$E_{\vec{k}} \sinh 2\alpha_{\vec{k}} = \left(\frac{N}{V}\right) V(\vec{k}) , \tag{4.5}$$

and for the spectrum one gets:

$$E_{\vec{k}} = \sqrt{\frac{\vec{k}^2}{m_4}\left(\frac{N}{V}\right) V(\vec{k}) + \frac{\vec{k}^4}{4m_4^2}} . \tag{4.6}$$

Going through the discussion that led to (1.94) and (1.95), we realize that (4.5) and (4.6) can only hold for momenta less than $k_0 \simeq \frac{2\pi}{a}$ ($a^3 = \frac{V}{N}$, in the case of ^{4}He $k_0 \simeq 1 \overset{\circ}{A}^{-1}$). For $|\vec{k}|$-values higher than k_0, $\alpha_{\vec{k}} \to 0$, and (4.6) yields

$$E_{\vec{k}} \simeq \frac{\vec{k}^2}{2m_4} + \left(\frac{N}{V}\right) V(\vec{k}) . \tag{4.7}$$

A careful analysis of this problem has been carried out with R. Mele and S. Villa[1] yielding the single particle spectrum, reported in Fig. 4.2

Leaving the 3- and 4-particle interactions aside, for they can clearly be treated perturbatively, we have thus obtained a remarkably simple description of the normal state of liquid ^{4}He: a simple gas of quasi-particle excitations, the "rotons",

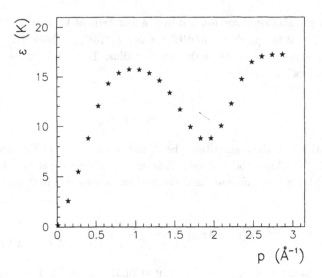

Fig. 4.2. The single-particle spectrum issuing from (4.6) with the "pseudopotential"(4.4)

with the peculiar spectrum (that was first guessed on thermodynamical grounds by Landau[2], and later completely confirmed by neutron and Raman scattering[3]) depicted in Fig. 4.2. In our QFT analysis the apparent kinematical and dynamical complications of a quantum liquid like ^{4}He are thus seen to disappear presenting us with a perfectly tame physical system, quite akin to a perfect gas! It goes without saying that the particles of this gas are **not** the ^{4}He-atoms, but new physical objects -the rotons- with an energy spectrum $\epsilon(q)$ (drawn in Fig. 4.2), whose deviation from the free particle spectrum clearly shows that it is the result of a non-trivial collective dynamics associated with the diagonalization of the Hamiltonian (4.2). One can go further and apply this model to the now straightforward statistical thermodynamics of the normal ^{4}He liquid. One must also note that there does not seem to be any particular reason why a similar analysis could not be carried out for all simple liquids.

4.2 "Weak coherence" in superfluid ^{4}He

As mentioned above, if at SVP we cool liquid ^{4}He below $T = T_\lambda = 2.15$ K, it stops being an almost perfect gas of "rotons" to become an inviscid, perfect fluid that flows in capillaries without friction.

Furthermore the disorder —entropy— created by a local heating in such a fluid does not propagate by diffusion, as all entropy perturbations normally do, but rather as waves that go at the relatively large speed of about 30 m/sec, the so called "second sound". As a result ^{4}He is a very calm fluid, no local boiling being allowed by the fast heat conduction associated with the "second sound". It was realized early on —in the thirties— that such properties could only belong to a macroscopic system that possesses a full quantum behaviour, *i.e.* is described by a single, macroscopic wave-function.

A large number of experiments have demonstrated, however, that the typical friction of a normal fluid reappears in ^{4}He when one puts into it, for instance, a winding mill in rapid motion. What happens then? Is Galilean invariance being violated by the different behaviours that the fluid exhibits when it is in motion with respect to the obstacle (the capillary walls) on one hand, and when the obstacle (the mill) is in motion with respect to the fluid on the other? This clearly does not make any sense. The resolution of this puzzle was discovered in the 40's by Landau[2], who postulated the existence of two fluids — the "super" (s) and the "normal" (n) — that interpenetrate each other, and such that the density $\rho(T)$ of ^{4}He as a function of temperature is given by

$$\rho(T) = \rho_s(T) + \rho_n(T) \; ; \tag{4.8}$$

where $\rho_s(T)$ and $\rho_n(T)$ are the densities of the superfluid and of the normal fluid respectively. At $T = 0$ $\rho_s(0) = \rho(0)$, while at $T_\lambda = 2.15$ K at SVP $\rho_n(T_\lambda) = \rho(T_\lambda)$.

The hypothesized existence of two fluids, while solving brilliantly the problem of Galilean invariance and explaining a large body of phenomenology, does not quite bring the "magic" of ^{4}He back to the realm of scientific rationalism. Indeed, let us look at liquid ^{4}He from the point of view of quantum physicists — like in the thirties Bernal and Fowler did for water — what do we see ? A very large crowd of identical compact billiard balls, subject to the quantum "smearing" dictated by Heisenberg principle held together by the very weak "hooks" of Van der Waals forces and prevented from condensing, in spite of their Bose character, by a very strong "hard-core". If this picture is correct, and nobody doubts it is correct, then a paradox emerges: how can these identical particles decide to which fluid they belong? Has Ockham's razor completely lost its sharpness? The paradox has been rendered much more severe by a series of neutron scattering measurements, that confirm the validity of the two-fluid picture even at the microscopic level. It is interesting that the incongruity of this situation is fully appreciated by CM physicists, I quote from a recent report of a well known expert in the field[4] "... after 50 years of study, superfluid ^{4}He remains in many ways much more mysterious than its younger relatives, superfluid ^{3}He and superconductivity". In this book we shall have something to say about these latter mysteries as well.

It will now be shown that QED coherence goes a long way toward solving the mysteries of ^{4}He-superfluidity. For this purpose[5] let us introduce the bosonic wave-field of ^{4}He,

$$\psi(\vec{x}, \alpha, t) = \sum_n a_n(t)\phi_n(\vec{x}, \alpha) , \qquad (4.9)$$

$\{\phi_n(\vec{x}, \alpha)\}$ being a complete system of the free atomic Hamiltonian, with $\vec{x}$ representing the center of mass and α the electronic coordinate. The wave-functions $\phi_n(\vec{x}, \alpha)$ thus correspond to the stationary energy levels of the ^{4}He atom. The CE's for the wave-field have been written down in their general form in (2.50) and (2.51), however they can be further simplified by retaining in the expansion (4.9) only the terms that participate in the coherent dynamics. In fact, in the initial incoherent configuration the only state that is populated is the atomic ground state $|1S\rangle$ of parahelium, and due to the validity of the dipole approximation the only atomic states that will be connected electromagnetically to the ground state are the states $|nP\rangle$ of parahelium. Thus, without any prejudice, we may write

$$\psi(\vec{x}, \alpha, t) = a_0(\vec{x}, t)\langle \alpha|1S\rangle + \sum_n \sum_{k=1}^{3} a_{nk}(\vec{x}, t)\langle \alpha|nP, k\rangle , \qquad (4.10)$$

and in terms of the amplitudes $a_0(t)$, $a_{nk}(t)$ and $\alpha_{\vec{k}r}(t)$ the CE's become (we neglect the very small kinetic energy term)

$$i\dot{a}_0(\vec{x}, t) = E_{1S}a_0(\vec{x}, t) + e\left(\frac{N}{V}\right)^{\frac{1}{2}} \sum_{n,\vec{k},r} \frac{1}{\sqrt{2\omega_{\vec{k}}}} \left[\alpha_{\vec{k}r}\vec{\epsilon}_{\vec{k}r}e^{-i(\omega t - \vec{k}\cdot\vec{x})} + \text{c.c.}\right]$$
$$\cdot \sum_k \langle 1S|\vec{j}_{em}|nP, k\rangle a_{nk}(\vec{x}, t), \qquad (4.11a)$$

$$i\dot{a}_{nk}(\vec{x}, t) = E_{nP}a_{nk}(\vec{x}, t) + e\left(\frac{N}{V}\right)^{\frac{1}{2}} \sum_{\vec{k},r} \frac{1}{\sqrt{2\omega_{\vec{k}}}} \left[\alpha_{\vec{k}r}\vec{\epsilon}_{\vec{k}r}e^{-i(\omega t - \vec{k}\cdot\vec{x})} + \text{c.c.}\right]$$
$$\cdot \langle nP, k|\vec{j}_{em}|1S\rangle a_0(\vec{x}, t), \qquad (4.11b)$$

$$i\dot{\alpha}_{\vec{k}r} - \frac{1}{2\omega_{\vec{k}}}\ddot{\alpha}_{\vec{k}r} = e\left(\frac{N}{V}\right)^{\frac{1}{2}} \frac{\vec{\epsilon}_{\vec{k}r}e^{i\omega_{\vec{k}}t}}{\sqrt{2\omega_{\vec{k}}}} \int_{\vec{x}} \left(\sum_{n,k} a_0^*(\vec{x}, t)a_{nk}(\vec{x}, t)\langle 1S|\vec{j}_{em}|nP, k\rangle + \text{c.c.}\right) e^{-i\vec{k}\cdot\vec{x}}. \qquad (4.11c)$$

Note that in (4.11c) we have neglected the "photon mass term", for all the important transitions from the ground state $|1S\rangle$ in the dipole approximation have been already taken into consideration. We now employ the "interaction representation", setting as usual

$$a_0(\vec{x}, t) = b_0(\vec{x}, t)e^{-iE_{1S}t}$$
$$a_{nk}(\vec{x}, t) = b_{nk}(\vec{x}, t)e^{-iE_{nP}t} ,$$

(4.12)

and obtain the new CE's ($\omega_n = E_{nP} - E_{1S}$)

$$i\dot{b}_0(\vec{x}, t) = e\left(\frac{N}{V}\right)^{\frac{1}{2}} \sum_{n,\vec{k},r} \frac{1}{\sqrt{2\omega_{\vec{k}}}} \left[\alpha_{\vec{k}r}\vec{\epsilon}_{\vec{k}r}e^{-i(\omega t - \vec{k}\cdot\vec{x})} + \text{c.c.}\right]$$
$$\cdot \sum_k \langle 1S|\vec{j}_{em}|nP, k\rangle e^{-i\omega_n t}b_{nk}(\vec{x}, t),$$

(4.13a)

$$i\dot{b}_{nk}(\vec{x}, t) = e\left(\frac{N}{V}\right)^{\frac{1}{2}} \sum_{\vec{k},r} \frac{1}{\sqrt{2\omega_{\vec{k}}}} \left[\alpha_{\vec{k}r}\vec{\epsilon}_{\vec{k}r}e^{-i(\omega t - \vec{k}\cdot\vec{x})} + \text{c.c.}\right] \cdot$$
$$\cdot \langle nP, k|\vec{j}_{em}|1S\rangle e^{i\omega_n t}b_0(\vec{x}, t),$$

(4.13b)

$$i\dot{\alpha}_{\vec{k}r} - \frac{1}{2\omega_{\vec{k}}}\ddot{\alpha}_{\vec{k}r} = e\left(\frac{N}{V}\right)^{\frac{1}{2}} \frac{1}{\sqrt{2\omega_{\vec{k}}}}\vec{\epsilon}^*_{\vec{k}r}e^{i\omega_{\vec{k}}t}$$
$$\cdot \int_{\vec{x}} \left(\sum_{n,k} b_0^*(\vec{x}, t)b_{nk}(\vec{x}, t)e^{-i\omega_n t}\langle 1S|\vec{j}_{em}|nP, k\rangle + \text{c.c.}\right)e^{-i\vec{k}\cdot\vec{x}}.$$

(4.13c)

Noting that all ω_n's do not differ by more than 10 % from the frequency of the 1S-2P transition $\omega = 21.2$ eV, we may consider them all coupled to the same e.m. plane wave with "free" frequency $\omega_{\vec{k}} = \omega$. We are now essentially in the same situation as in the case of the two-level system, and following the same strategy of Section (3.2) we get the space-independent equations (valid within the coherence domain of size $\lambda = \frac{2\pi}{\omega} \simeq 600\text{Å}$)

$$\dot{b}_0(\tau) = \sum_{n,k} g_n\alpha_k^* b_{nk}(\tau)$$

$$\dot{b}_{nk}(\tau) = -g_n\alpha_k b_0(\tau)$$

(4.14)

$$\dot{\alpha}_k(\tau) + \frac{i}{2}\ddot{\alpha}_k(\tau) = \sum_n g_n b_0^*(\tau)b_{nk}(\tau) ,$$

where $\tau = \omega t$,

$$\alpha_k(\tau) = \sqrt{\frac{3}{4\pi}} \sum_r \int d\Omega_{\vec{k}}\alpha_{\vec{k}r}\vec{\epsilon}_{\vec{k}r} ,$$

(4.15)

$$g_n = \sqrt{\frac{2\pi}{3}}\lambda_n\left(\frac{\omega_p}{\omega}\right) \ , \tag{4.16}$$

where the classical plasma frequency ω_p has been introduced (m_e is the electron mass) $\omega_p = e\left(\frac{N}{V}\right)^{\frac{1}{2}}\frac{1}{\sqrt{m_e}}$, and λ_n parametrises the matrix element

$$\langle nP, k|j_i|1S\rangle = \langle nP, k|\frac{P_i}{m_e}|1S\rangle = -i\lambda_n\sqrt{\frac{\omega_n}{2m_e}}\delta_{ik} \ . \tag{4.17}$$

The two constants of motion of the system (4.14) are

$$|b_0(\tau)|^2 + \sum_{n,k}|b_{nk}(\tau)|^2 = 1 \ , \tag{4.18}$$

and

$$Q = \sum_k\left\{|\alpha_k(\tau)|^2 + \frac{i}{2}[\alpha_k^*(\tau)\dot{\alpha}_k(\tau) - \dot{\alpha}_k^*(\tau)\alpha_k(\tau)] + \sum_n|b_{nk}(\tau)|^2\right\} \ , \tag{4.19}$$

whose meanings are precisely the same as those of their two-level analogues (3.13) and (3.14). The short-time dynamics of the e.m. field amplitudes is now governed by the linear differential equation

$$\ddot{\alpha}_k(\tau) + \frac{i}{2}\dddot{\alpha}_k(\tau) = -g^2\alpha_k(\tau) \ , \tag{4.20}$$

where the coupling constant g^2 is given by

$$g^2 = \sum_n g_n^2 = \frac{2\pi}{3}\left(\frac{\omega_p}{\omega_0}\right)^2\sum_n\lambda_n^2 \ . \tag{4.21}$$

By substituting the experimental values at zero temperature and SVP one computes $g \simeq .25$.[5] But we know from (3.19) that the threshold for strong coherence is at

$$g = g_c = \sqrt{\frac{16}{27}} = .77. \tag{4.22}$$

Thus ^{4}He is only "weakly coherent". However, as discussed in Chapter 3, for systems whose coherence domains are quite small (as in our case) weak coherence may still have noticeable effects. Note that at $T = 0$ and at SVP there are about $N_{CD} = 3\cdot 10^6$ atoms in the CD.

Let us now study the stationary configuration corresponding to the Coherent Ground State (CGS). We set as usual:

$$b_0(\tau) = B_0 e^{i\psi_0(\tau)}$$
$$b_{nk}(\tau) = B_n {}_k e^{i\psi_k(\tau)} \qquad\qquad (4.23)$$
$$\alpha_k(\tau) = A_k e^{i\phi_k(\tau)} .$$

We note at this point that, due to the rotational invariance of (4.14), the solutions of our system have a basic degeneracy, which can be labelled by a unit complex vector u_k ($\vec{u}^* \cdot \vec{u} = 1$). Thus in terms of such parameters the above equations can be rewritten as

$$b_0(\tau) = B_0 e^{i\psi_0(\tau)}$$
$$b_{nk}(\tau) = B_n u_k e^{i\psi_1(\tau)} \qquad\qquad (4.24)$$
$$\alpha_k(\tau) = A u_k e^{i\phi(\tau)} .$$

The normalization (4.18) demands:

$$B_0{}^2 + \sum_n B_n{}^2 = 1 , \qquad\qquad (4.25)$$

which suggests that we set $B_0^2 = \cos^2 \theta$, $\sum_n B_n^2 = \sin^2 \theta \simeq \theta^2 = \sum_n \theta_n^2$, in view the fact that $\cos \theta \simeq 1$, which always holds in a situation of "weak coherence".

Substituting (4.24) in (4.14) we obtain:

$$\dot{\psi}_0 = A \sum_n g_n \theta_n = gA\theta, \qquad\qquad (4.26a)$$

$$\dot{\psi}_1 = A \frac{g_n}{\theta_n} = \frac{gA}{\theta}, \qquad\qquad (4.26b)$$

$$\dot{\phi} - \frac{1}{2}\dot{\phi}^2 = \sum_n \frac{g_n}{6\pi} \frac{\theta_n}{A} = \frac{g\theta}{6\pi A} , \qquad\qquad (4.26c)$$

where we have set $\theta_n = \frac{g_n}{g}\theta$, as demanded by the structure of the system. In terms of $A, \dot{\phi}$ and θ the conserved quantity (4.19) becomes

$$Q = A^2(1 - \dot{\phi}) + \theta^2. \qquad\qquad (4.27)$$

The value of Q can be computed by substituting in (4.19) for the e.m. amplitude the initial conditions ($\dot{\alpha}_k = 0$),

$$\sum_k |\alpha_k(0)|^2 = |A|^2 = \frac{4\pi \cdot 2}{2N_{CD}} , \qquad\qquad (4.28)$$

corresponding to an intensity $\frac{1}{2N_{CD}}$ for each of the $2 \cdot 4\pi$ independent degrees of freedom of the free e.m. field, and for the bosonic matter field

$$\theta^2(0) = \frac{4\pi \cdot 3}{2N_{CD}} , \qquad (4.29)$$

as there are $3 \cdot 4\pi$ independent degrees of freedom in the p-wave modes. Note that, according to (3.15), Q is nothing but the energy (divided by ωN_{CD}) contained in the zero-point fluctuations of both matter and e.m. fields. Substituting in (4.27) one gets

$$Q = \frac{10\pi}{N_{CD}} . \qquad (4.30)$$

Eqs. (4.26), (4.27) and (4.30), together with the condition

$$\psi_0 - \psi_1 + \phi = -\frac{\pi}{2} , \qquad (4.31)$$

characterize completely the CGS solution. Indeed substituting the solutions of (4.26) in the time derivative of (4.31) we obtain:

$$1 - \sqrt{1 - \frac{2g\theta}{A}} = \frac{gA}{\theta} , \qquad (4.32)$$

while from (4.27) we can write

$$A^2\sqrt{1 - \frac{2g\theta}{A}} + \theta^2 = \frac{10\pi}{N_{CD}} , \qquad (4.33)$$

which completely determine A, θ and through (4.26) all other relevant quantities. Evidently u_k remains completely undetermined.

Let us now compute the energy difference ΔE between the CGS and the Perturbative Ground State (PGS): the gap δ, $i.e.$ the energy that one needs in order to remove a particle from the CGS and send it to fluctuate around the PGS, obviously equals $\frac{-\Delta E}{N}$. We have

$$\frac{\Delta E}{N} = \frac{E_{matt}}{N} + \frac{E_{em}}{N} + \frac{E_{int}}{N} - \frac{E_{PGS}}{N} , \qquad (4.34)$$

where

$$\frac{E_{matt}}{N} = \omega\theta^2 , \qquad (4.35)$$

a positive term that comes from the excitation of the matter P-wave modes;

$$\frac{E_{em}}{N} = \omega A^2 \cdot \frac{1}{2} \left[\left(\left(1 - \dot{\phi} \right)^2 + 1 \right) \right] = \omega A^2 \left[1 - \frac{g\theta}{A} \right] , \qquad (4.36)$$

is the e.m. field energy that one computes for the CGS from the expression (2.38b);

$$\frac{E_{int}}{N} = -2g\omega A\theta , \qquad (4.37)$$

and finally

$$\frac{E_{PGS}}{N} = \omega \left[\theta(0) \right]^2 + \frac{4\pi\omega}{N} . \qquad (4.38)$$

Putting numbers in [Ex. $\langle 4.2 \rangle$] we obtain for the gap δ

$$\delta \simeq 1 \mathrm{K} , \qquad (4.39)$$

a small but not negligible contribution, if we consider that the binding energy of each particle in the liquid is of the order of 5 K.

Another remarkable result is that the renormalized frequency ω_r of the e.m. field is

$$\omega_r = \left(1 - \dot{\phi} \right) \omega = \omega \sqrt{1 - \frac{2g\theta}{A}} < \omega , \qquad (4.40)$$

implying that the coherent e.m. field gets trapped spontaneously inside the condensed ^{4}He phase. In view of the smallness of θ we can conclude that at $T = 0$ the coherent electrodynamical interaction has generated a condensed phase of ^{4}He characterized by an order parameter which is nothing but the complex scalar field $b_0(\vec{x}, t)$.

Recalling that the macroscopic phenomenology of a superfluid is completely accounted for by the existence of such a physical object, we may conclude that we have found a simple *raison d'être* of ^{4}He superfluidity.

If we now slowly increase the temperature T, the discussion upon the two-fluid picture, carried out in Section (2.4), instructs us that a normal fluid of phonons and "rotons", with density $\rho_n(T)$, will build up first at the boundaries of CD's and then within the CD themselves. Following Landau[9] the density $\rho_n(T)$ is given by

$$\frac{\rho_n}{\rho} \simeq \left(\frac{\rho_n}{\rho} \right)_{rot} = \frac{1}{3\rho T} \int \frac{d^3 p}{(2\pi)^3} e^{-\frac{\epsilon(p,\rho,T)}{T}} p^2 , \qquad (4.41)$$

$\epsilon(p, \rho, T)$ being the single particle spectrum whose shape appears in Fig. 4.2. It turns out that $\epsilon(p, \rho, T)$ experimentally depends on both density and temperature; in particular, parametrising the roton minimum as

$$\epsilon(p, \rho, T) = \Delta(\rho, T) + \frac{(p - p_0(\rho, T))^2}{2\mu} \, , \tag{4.42}$$

the "gap" Δ depends on both density and temperature. How can one understand a dependence of the gap on temperature? The answer is quite simple: when the fraction $\frac{\rho_n}{\rho}$ increases, $\frac{\rho_s}{\rho} = 1 - \frac{\rho_n}{\rho}$ decreases, resulting in a renormalization of the coupling $g^2(\rho)$ and of the amplitudes B_0^2 and θ^2 by the factor $\left(\frac{\rho_s}{\rho}\right)$, leading to a temperature-dependent gap

$$\delta(T) = \delta(0)\frac{\rho_s}{\rho}(T) = \delta(0)\left[1 - \frac{\rho_n}{\rho}\right] \qquad (\delta(0) = \delta) \, . \tag{4.43}$$

At $\frac{\rho_n}{\rho} = 1$ the normal fluid will have invaded all ^{4}He liquid, and coherence, hence superfluidity is gone: we have thus the superfluid-normal fluid phase transition, and a vanishing gap $\delta(T_c)$; the phase transition is, as experimentally observed, of second order. This discussion also implies that

$$\Delta(\rho, T) = \Delta(\rho, T_c) + \delta(0)\,\frac{\rho_s}{\rho}(T) = \Delta(\rho, 0) - \delta(0)\,\frac{\rho_n}{\rho}(T) \, , \tag{4.44}$$

for at $T = 0$ the "gap" $\Delta(\rho, T_c)$ that is associated with the roton spectrum in the normal fluid is increased by $\delta(0)$, the energy that we must spend to "break" the coherence of the CGS. How well this simple picture fits the data for the quantities $\Delta(0) - \Delta(T)$ and $\frac{\rho_n}{\rho}$ can be glimpsed in the Fig. 4.3 and Fig. 4.4 respectively.

But the most satisfying side of the two-fluid picture, that as we have argued in Section (2.4), is characteristic of e.m. coherence in matter, is its elegant solution of the problems with Ockham's razor, alluded to above. To the question: how can the identical atoms decide to which fluid they belong? the answer is really simple: if they are in phase with the e.m. condensate, they belong to the CGS making up the superfluid, if on the other hand they have been excited incoherently, they fluctuate around the PGS and they have all the features of the quasi-particles of the normal fluid. Finally, one of the puzzles of ^{4}He superfluidity is by what mechanism superfluidity is lost when the superfluid flows in the capillaries with a velocity $v > v_{crit}$. It was again Landau[9] that showed that v_{crit} must be equal to the minimum value of $\frac{\epsilon(p)}{p}$, where $\epsilon(p)$ is the energy of a generic excitation carrying momentum p. Were the relevant excitations the single particle excitations one would obtain $v_{crit} \simeq 50$ m/sec, two orders of magnitude larger than the observed $v_{crit} \simeq 50$ cm/sec. This means that some other type of excitations must contribute to the loss of superfluidity in such a physical situation. These were correctly identified by Feynman and Onsager[10] as the

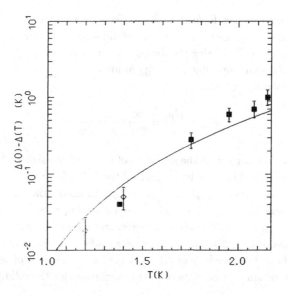

Fig. 4.3. The gap vs. temperature compared with experimental data obtained by Mezei[6] ($\square$) and Woods and Svensson[7] ($\diamond$).

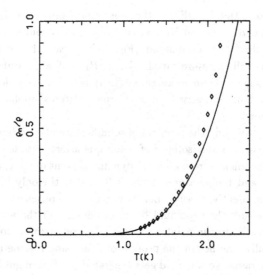

Fig. 4.4. ρ_n vs. temperature compared with experimental data[8]

quantized vortices that get formed in the perfect, irrotational superfluid. To them there correspond rotating motions of the superfluid around the vortex singularity with average momenta $p \simeq \frac{2\pi}{d}$, d being the dimensions of vortex loops (for vortices usually have the form of closed loops, due to energy minimization). Thus for such excitations one would have approximately

$$v_{crit} = \frac{\epsilon(p)}{p} = \frac{p^2}{2m} \cdot \frac{1}{p} = \frac{p}{2m} \; . \tag{4.45}$$

It is a prediction following from the existence of CD's (whose size for ^{4}He is $\lambda = \frac{2\pi}{\omega} = 600 \mathring{A}$) that vortices will form at boundaries of the CD's, for it will cost no energy to produce the vortex singularity there being no gap to access the normal state. Thus the maximum possible v_{crit} can be estimated by setting $p \simeq \frac{2\pi}{\lambda} = \omega \simeq 20$ eV in (4.45). One obtains $v_{crit} \simeq 70$ cm/sec, in agreement with observation, thus finding a nice confirmation of the existence of CD's. We could go on for quite a while in our tour through the phenomenology of ^{4}He, but this is not the aim of this book, so we will turn our attention to a possible coherent mechanism for the solidification of ^{4}He.

4.3 Beyond the Born-Oppenheimer approximation: the solidification of ^{4}He

We have just seen that by QFT methods one can obtain a simple and satisfactory picture of normal and superfluid ^{4}He. What can one say about solid ^{4}He?

By looking at the ^{4}He phase diagram (Fig. 4.1) we learn that at zero temperature the solid forms past the pressure threshold $P = P_{th} = 25$ atm. But solid ^{4}He, whose lattice is of the hexagonal close packing type (hcp), is rather anomalous, characterized by a large molar volume and very large non-harmonic atoms' oscillations around their equilibrium positions.

The possibility to study the problem of solidification at zero temperature[11], that in nature only occurs for the isotopes of helium, is a fortunate one, for it allows us to look at this phenomenon from a purely dynamical point of view, unclouded by the effects of entropy and temperature. Thus at T = 0 K the only important variable is pressure or, equivalently, density. Let us visualize the physics of solidification: by increasing the pressure the superfluid becomes denser and the average interatomic distance r smaller. From the point of view of electrodynamical coherence we must watch very carefully whether in the process of decreasing r some coupling between the atoms and the radiative e.m. field gets generated, for this might be the germ from which strong e.m. coherence will eventually grow, possibly producing the observed liquid-solid phase transition. It is clear that our attitude here is completely "orthogo-

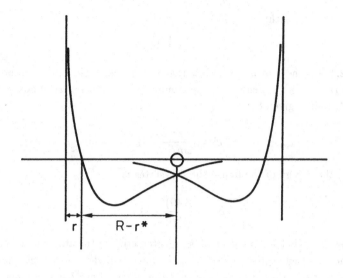

Fig. 4.5. Average potential for a ^{4}He atom

nal " to the GACMP view, for which the liquid-solid transition is to be attributed to some subtle, almost impossible to pin down precisely, configurational energy difference between the solid and the liquid, stemming from short-range electrostatic forces. In fact I think that to base such an outstanding phase transition purely on tiny quantitative details without introducing some dramatic qualitatively different interaction is hardly believable.

We need not look too far to identify a good candidate mechanism in the failure of the Born-Oppenheimer approximation, that occurs when the interatomic distance decreases towards the atomic size. According to the Born-Oppenheimer approximation the dynamics of the atomic nucleus does not differ from that of the stationary neutral atom or molecule to which it belongs, for during the typical times of its evolution the much faster dynamics of the electrons will average out any non-stationary configuration. Thus insofar as this approximation is applicable, the ^{4}He nucleus is effectively neutral, decoupled from any e.m. field mode. However let us now look at the local motion of the single atom in the liquid phase. Due to the interaction with its nearest neighbours, following F. London[12], we may visualize it as confined in the "cage" built around it by the neighbouring atoms. Summing over the twelve nearest neighbours of a hcp structure, the average potential that a ^{4}He atom experiences is depicted in Fig. 4.5.

For simplicity we may approximate this potential with that of a spherical box of radius $R = r - r^*$, with:

$$r = \rho^{-\frac{1}{3}} 2^{\frac{1}{6}} , \tag{4.46}$$

and $r^* = 2.4$ Å is the "covolume". Note that r^* has been estimated by imposing (See Fig. 4.5) $V(r^*) = E_0$, the energy of the ground state in the spherical box with radius $R = r - r^*$, that is given by:

$$E_0 = \frac{\pi^2}{2mR^2} - V_0 . \tag{4.47}$$

Recall that the energy of the first excited P-states is:

$$E_1 = \frac{(4.49)^2}{2mR^2} - V_0 . \tag{4.48}$$

It is now clear that when r decreases the interaction of the atoms with the "walls" of the spherical cage —the nearest neighbours— will deform the electronic shell around the nucleus, thus leading to a violation of the Born-Oppenheimer approximation: due to both the Pauli principle and the electrostatic repulsion the electrons cannot follow completely the motion of the nucleus, and a well defined average electric dipole moment gets thus generated. Can we estimate its size? Fortunately the repulsive part $\Delta(r_0/r)^{12}$, [with $\Delta = 10.22$ K and $r_0 = 2.869$Å] of the Lennard-Jones potential provides this estimate, for we can set ($\alpha = e^2/4\pi$ is the fine-structure constant):

$$\Delta(\frac{r_0}{r})^{12} \simeq 4\alpha \frac{D^2(r)}{r^3}, \tag{4.49}$$

thus expressing the short-range repulsion as the electrostatic interaction energy of two dipoles of strength $2eD(r)$, polarized in opposite directions. Note that in deriving (4.49) we have assumed that the origin of short-range repulsion is merely electrostatic and is due to the electric dipoles arising from the deformation of the electrons' distribution in the atoms, that depends very strongly on the relative atomic distance. In view of there being nothing but electrostatic forces acting at short distance, such assumption must be regarded as completely natural. The strong r-dependence of $D(r)$, that can be read in (4.49), leads us to expect a very strong variation of the e.m. coupling with interatomic distance: precisely what we are looking for!

We possess now all the ingredients needed to analyse a possible mechanism for explaining the liquid-solid phase transition as due to the possible emergence of strong electrodynamical coherence. Indeed the matter system now comprises a set of atoms performing oscillations inside spherical cages, whose stationary states have wavefunctions $|0\rangle$ and $|1k\rangle$ with frequencies (4.47) and (4.48) respectively. By a procedure that should be familiar by now the relevant wave-field is:

$$\Psi(\vec{x}, \vec{r}, t) = b_o(\vec{x}, t) \langle \vec{r}|0\rangle + b_k(\vec{x}, t)\langle \vec{r}|1k\rangle, \qquad (4.50)$$

and the frequency of the coupled e.m. field is accordingly:

$$\omega(r) = E_1 - E_0 = \frac{10.29}{2m(r - r^*)^2}, \qquad (4.51)$$

which for a typical liquid density ($r \simeq 4\overset{\circ}{A}$) is equal to 106 cm^{-1}, implying a size of the CD's $\lambda = \dfrac{2\pi}{\omega(r)} \simeq 10^{-2}$cm, a rather large value.

The interaction Hamiltonian between the e.m. field and the atomic electric dipole $2e\vec{D}$

$$H_{int} = -e\vec{E} \cdot \vec{D}, \qquad (4.52)$$

takes the QFT form:

$$H_{int} = -e \int_V d^3\vec{x}\, d^3\vec{r}\, \vec{E}(\vec{x}, t)\Psi^\dagger(\vec{x}, \vec{r}, t) \cdot \vec{D}\Psi(\vec{x}, \vec{r}, t)$$

$$= -e \int_V d^3\vec{x} \sum_{\vec{k}\, r} i\left(\frac{\omega_{\vec{k}}}{2V}\right)^{1/2} [\alpha_{\vec{k}r}e^{-i(\omega_{\vec{k}}t - \vec{k}\cdot\vec{x})}\vec{\epsilon}_{\vec{k}r} - \text{c.c.}] \cdot \qquad (4.53)$$

$$\cdot [b_0^*(\vec{x}, t)b_k(\vec{x}, t)\langle 0|\vec{D}|1k\rangle + \text{c.c.}],$$

where the transition dipole matrix element is:

$$\langle 0|D_i|1k\rangle \simeq \delta_{ik}\bar{D}(r) \qquad (4.54)$$

with

$$\bar{D} = \frac{1}{\sqrt{3}} \int \rho^2 d\rho\, \phi_0^*(\rho)D(\rho)\phi_1(\rho), \qquad (4.55)$$

$\phi_0 \simeq j_0(\pi\rho/R)$ and $\phi_1 \simeq j_1(4.49\rho/R)$ are the normalized spatial eigenfunctions of the spherical cavity of radius $R = r - r^*$. A good approximation for $\bar{D}$ can be obtained as follows [Ex. $\langle 4.3\rangle$]

$$\bar{D} = \frac{D(r)}{\sqrt{3}} \int_0^R \rho^2 d\rho \frac{\phi_0^*(\rho)\phi_1(\rho)}{[1 - \rho/r]^{9/2}} \simeq \frac{D(r)}{\sqrt{3}} \frac{1}{[1 - \bar{\rho}/r]^{9/2}} \qquad (4.56)$$

where $\bar{\rho} \simeq .39(R)$ is the transition radius $\langle 0|r|1\rangle$.

Without any further ado, following the steps described in the preceding Section, we may write the space-independent CE's as ($\tau = \omega t$, as usual):

$$\dot{b}_0(\tau) = g\alpha_k^*(\tau)b_k(\tau) , \tag{4.57a}$$

$$\dot{b}_k(\tau) = -g\alpha_k(\tau)b_0(\tau) , \tag{4.57b}$$

$$\dot{\alpha}_k(\tau) + \frac{i}{2}\ddot{\alpha}_k(\tau) = gb_0^*(\tau)b_k(\tau) , \tag{4.57c}$$

where

$$\alpha_j(t) = \sqrt{\frac{3}{8\pi}}\sum_r \int d\Omega_k \alpha_{\vec{k}r}(\tau)\epsilon_{kr,j}, \tag{4.58}$$

and the coupling constant

$$g(r) = \sqrt{\frac{4\pi}{3}}\frac{2e\bar{D}(r)}{\sqrt{\omega(r)}}\sqrt{\frac{N}{V}}, \tag{4.59}$$

depends strongly on the interatomic distance r through its dependence on $D(r)$, $\omega(r)$ and $\rho = \dfrac{N}{V}$. For instance at $P = 0$ atm, $r(0) = 4$ Å we compute

$$g(0) = 0.69 , \tag{4.60}$$

while at P=25 atm experimentally one has $r(25) = 3.79$ Å , and we get instead

$$g(25) = 0.78 . \tag{4.61}$$

Eqs. (4.60) and (4.61) are extremely interesting for we know that for the system (4.57) the critical coupling constant g_c for the spontaneous establishment of strong coherence, hence of a non-trivial macroscopic e.m. condensate, is just:

$$g_c = \sqrt{\frac{16}{27}} = 0.77. \tag{4.62}$$

Our simple analysis thus shows that at about P=25 atm something special happens, the superfluid liquid "runs away" to a new physical system whose structure must display the basic spatial features of the solid state.

But before we analyse this fascinating problem, let us discuss the features of the solution of the differential system in the neighborhood of $g = g_c$. Setting as usual

(u_k is the familiar complex vector that labels the degenerate solutions)

$$b_o(\tau) = \cos\theta e^{i\psi_0(\tau)} ,\qquad\qquad (4.63)$$

$$b_k(\tau) = u_k \sin\theta e^{i\psi_1(\tau)} ,\qquad\qquad (4.64)$$

$$\alpha_k(\tau) = A u_k e^{i\phi(\tau)} ,\qquad\qquad (4.65)$$

$$(4.66)$$

and carrying out the customary calculation [Ex. $\langle 4.4\rangle$] (see the preceding Section) the energy difference ΔE between the coherent state of minimum energy (the CGS) and the PGS turns out to be: $[b_0 = 1 , \; |b_k| \simeq \frac{1}{\sqrt{N_{cd}}} , \; N_{cd} \simeq 10^{18}]$

$$\delta = \frac{\Delta E}{N} = \omega \left[\frac{1 - 3\epsilon}{2(1 - 2\epsilon)} + \frac{g^2}{2} \frac{3\epsilon^2 - 1}{(1 - \epsilon^2)^2} \cdot \frac{1 + 3\epsilon^2 - 4\epsilon}{(1 - 2\epsilon)^2} \right] \qquad (4.67)$$

where

$$\epsilon = \sqrt{1 - \frac{g \sin 2\theta}{A}}. \qquad\qquad (4.68)$$

One can now set:

$$\epsilon = \frac{1}{3}(1 - \gamma); \qquad\qquad (4.69)$$

in the neighborhood of $g \simeq g_c$ γ is a small quantity. To first order in γ the "gap" δ (4.67) can be written:

$$\delta \simeq -\frac{(g^2 - g_c^2)^2}{2g^2 g_c^2} \omega \qquad\qquad (4.70)$$

exhibiting the exothermal nature of the transition to the CGS.

One can solve the system (4.57) with the usual "momentum conservation" equation numerically for any value of g. The behaviour of the gap δ as a function of g appears in Fig. 4.6, which shows that by increasing $g(r)$ (according to (4.49) and (4.59) this is achieved by decreasing the interatomic distance r) one gains a substantial amount of energy. But evidently this is not the whole story, for the short-range interaction plays a fundamental rôle in determining the equilibrium density (through the repulsive core terms) and possibly the specific lattice structure. Nevertheless invoking Ockham's razor the very emergence of a lattice structure must be fundamentally related with the kind of coherent phase transition that the liquid undergoes as a result of the failure of the Born-Oppenheimer approximation. Let us see how this comes possibly about.

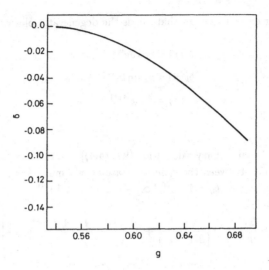

Fig. 4.6. Gap vs coupling constant

When the liquid "runs away", having reached through external compression the critical coupling g_c, it will continue for a while to get squeezed, but now "spontaneously" for this is what the behaviour of the gap reported in Fig. 4.6 is telling us. At constant pressure this squeezing will stop when the e.m. energy gain will be compensated by the energy increase due to the short-range repulsive interaction,which may become very strong. In so doing the system changes discontinuously (over the macroscopic times of our observation) its spatial structure: the phase transition is, as observed, first order. But why is this structure different from a simply denser type of liquid? Why do the atoms lose their freedom to move around unhindered?

Remarkably the answer to this question, that has been usually "answered pragmatically",*i.e.* by merely invoking the physical reality of crystals, is in the very nature of the coherent process that involves the ^{4}He nuclei in a dynamical situation where the Born-Oppenheimer approximation fails. These nuclei emit and absorb e.m. radiation by means of electric dipoles instead of the usually negligible magnetic moments. We have seen above that the frequency ω of the "resonating" e.m. waves is determined by the local space-structure of the ensemble of atoms, through the "cage" that must establish itself if the individual nucleus is to get coupled to the e.m. field. In the liquid this "cage" appears as a short transient phenomenon, for there is no energy to be gained (weak coherence is an extremely small effect) by its getting stabilized. But above g_c the system becomes strongly coherent and, as Fig. 4.6 shows, there is quite a large energy to be gained by maintaining $\omega(r)$ fixed, and in tune with the e.m. condensate. This latter requirement demands that the average (equilibrium) positions of the atoms stay fixed in space: the atoms **must** form a lattice. In order to remove an atom from such fixed position we must not only pay the gap δ for the atom but also for the atoms that get disturbed by such a displacement; and this is in general very large unless we are close to the boundaries of the CD (whose size $\lambda = \dfrac{2\pi}{\omega} \simeq 600\mu$, as we have noticed, is very large). We leave here the discussion of the connection between coherent e.m. processes involving the nuclei and the phenomenological aspect of solids and crystals; we shall resume it in Chapter 5.

To end this Section the energetics of the liquid-solid transition shall be briefly discussed. As we have noticed, at P=25 atm, when strong coherence sets in, the ensemble of ^{4}He atoms decrease their molar volume until the enthalpies for the liquid and the solid become equal. This condition can be expressed as follows:

$$U_{0l}(r_l) + Pv_l + \tilde{\delta}(r_l) = U_{0s}(r_s) + Pv_s + \tilde{\delta}(r_s) + \delta \tag{4.71}$$

where r_s and r_l are the interatomic distances for the solid and the liquid respectively, v_s and v_l their molar volumes, $\tilde{\delta}$ is the contribution from the coherent electronic process discussed in the preceding Section that depends very weakly on density, δ is the gap due to the coherence of nuclei and, finally, $U_{0s}(r_s)$ and $U_{0l}(r_l)$ are the energy contributions from short-range forces in the two phases. It is clear that, according to our view, $U_0(r)$ must be a universal function solely dependent on the interatomic distance and not on space order, the energy difference between liquid and solid being mainly due to the setting in of strong QED coherence. From a rather refined calculation[13], utilizing the interatomic distance r_s and r_l at the transition for the solid and the liquid respectively, we have:

$$U_0(r_s) = -5.5 \text{ K/part}, \tag{4.72}$$

$$U_0(r_l) = -6.4 \text{ K/part}. \tag{4.73}$$

On the other hand from (4.67) we compute:

$$\delta = -0.4 \text{ K/part}, \tag{4.74}$$

from which we obtain (see (4.71)):

$$P(v_l - v_s) = U_0(r_s) - U_0(r_l) + \delta = 0.5 \text{ K/part}, \tag{4.75}$$

to be compared with the experimentally observed value:

$$P(v_l - v_s)|_{\exp} = 0.6 \text{ K/part}, \tag{4.76}$$

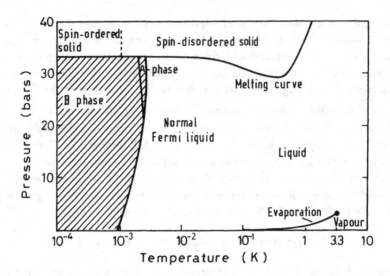

Fig. 4.7. Superfluid ^{3}He: Phase Diagram

a remarkable agreement.

We could now extend this analysis to non-zero temperatures, but again this book is not a treatise, so we shall leave solid ^{4}He at that.

4.4   Electrodynamical coherence in an ensemble of fermions: a clue to the superfluidity of ^{3}He

Even though at the atomic level ^{3}He shows very little difference from the more common ^{4}He, when part of a macroscopic ensemble ^{3}He behaves in a completely different manner, as witnessed by the P-T phase diagram reported in Fig. 4.7. Looking at the phase diagram we notice that also ^{3}He remains liquid down to zero temperature, but the superfluid, which in absence of an external magnetic field comprises two different phases **A** and **B**, only forms at T=1 mK, a temperature one thousand times smaller than in ^{4}He.

Qualitatively, the large difference in the collective behaviours of the two helium isotopes is well understood (or, at least it is believed to be well understood): one is a bosonic system while the other consists of fermions, obeying the Pauli principle. The configuration space of a fermion wave-field is in fact completely different from that of a bosonic field, and this is bound to have important consequences on its dynamics. In order to understand this, following the ideas of Chapter 1, let us introduce the fermionic wave-field:

$$\Psi(\vec{x}, t) = \sum_{\vec{k},\alpha} a_{\vec{k}\alpha}(t)\chi_{\vec{k}\alpha}(\vec{x}),\tag{4.77}$$

where $\chi_{\vec{k}\alpha}$ is a complete set of eigenfunction of the free Hamiltonian and α is the internal variable, usually comprising a dichotomic spin variable as in the case of ^{4}He. The quantum conditions have the form:

$$\{a_{\vec{k}\alpha}, a_{\vec{k}'\alpha'}^\dagger\} = \delta_{\vec{k}\vec{k}'}\delta_{\alpha\alpha'}.\tag{4.78}$$

In the free case, i.e. when the Hamiltonian has the form:

$$H = H_0 + \sum_{\vec{k}\alpha} \epsilon(\vec{k})a_{\vec{k}\alpha}^\dagger a_{\vec{k}\alpha},\tag{4.79}$$

the ground state $|0\rangle$ (which, modulo perturbations, coincides with the PGS) has a very simple structure:

$$|0\rangle = \prod_\alpha |0\rangle_\alpha\tag{4.80}$$

where for $\alpha = s$, the ^{3}He ground state, we have:

$$|0\rangle_s = \prod_{|\vec{k}|\le k_F} |1,\uparrow\rangle_{\vec{k}s}\, |1,\downarrow\rangle_{\vec{k}s} \prod_{|\vec{k}|>k_F} |0,\uparrow\rangle_{\vec{k}s}|0,\downarrow\rangle_{\vec{k}s},\tag{4.81}$$

and for the excited atomic state with $\alpha \ne s$:

$$|0\rangle_\alpha = \prod_{\vec{k}} |1,\uparrow\rangle_{\vec{k}\alpha}\, |1,\downarrow\rangle_{\vec{k}\alpha}\tag{4.82}$$

where $|0,1;\uparrow,\downarrow\rangle_{\vec{k}\alpha}$ represent either the unoccupied state or the occupied state for each spin of the plane wave of momentum $\vec{k}$, and k_F is the Fermi momentum defined by

$$N = 2\sum_{|\vec{k}|\le k_F};\tag{4.83}$$

The meaning of (4.80) is very simple: in the PGS all s-states are occupied up to the Fermi momentum, above such momentum they are empty and are represented by the product of the ground states of all Fermi oscillators with $|k| > k_F$, as for all other states they are unoccupied. The structure of the PGS (4.80) at T= 0 K is to be contrasted with that of the PGS of a Bose system where the lowest state $(\vec{k} = 0)$ is occupied by all the N particles. One of the typical features of the PGS of a fermionic system is that around the Fermi surface particles can be excited with a spectrum given by :

$$\epsilon(\vec{k} - \vec{k}_F) = \frac{k^2}{2m} - \frac{k_F^2}{2m} = \frac{(\vec{k} + \vec{k}_F)}{2m} \cdot (\vec{k} - \vec{k}_F) \simeq v_F |\vec{k} - \vec{k}_F|, \qquad (4.84)$$

where $v_F = \dfrac{k_F}{m}$ is the Fermi velocity.

Let us get back to ^{3}He: its wave-field takes up a form of the type (4.77), where $(s = +, -)$ is the ^{3}He spin nuclear variable; (please note the simplified notation):

$$\chi_{\vec{k}\alpha}(\vec{x}, \vec{\xi}) = \frac{e^{i\vec{k}\cdot\vec{x}}}{\sqrt{V}}[\langle \vec{\xi}|S, s\rangle, \ \langle \vec{\xi}|Pk, s\rangle], \qquad (4.85)$$

and like in the case of ^{4}He the states $|S\rangle$ and $|Pk\rangle$ refer to the well known eigenstates of parahelium.

The PGS is of the form (4.80), and the Fermi sphere comprises all atoms in the atomic ground state $|S\rangle$; all other states are clearly empty. We expect however that coupling our fermions to the e.m. field will lead to some transition from the S-state to the P-states, thus we look for a solution of the CE's, that for a fermionic system have the form (2.53) and (2.54), where for the ground state $|\Omega, t\rangle$ we set ($\sum_k |u_k|^2 = 1$):

$$|\Omega, t\rangle = \prod_{\vec{p}, s}(\cos \alpha_{\vec{p}}(t) + u_k \sin \alpha_{\vec{p}}(t) a^\dagger_{p\ Pk, s} a_{p\ S, s}) \, |O\rangle \ . \qquad (4.86)$$

The space independent Schrödinger equation reads:

$$i\frac{\partial}{\partial t}|\Omega, t\rangle = \sum_{ps}(E_{1S} a^\dagger_{p\ S, s} a_{p\ S, s} + E_{2P} a^\dagger_{p\ Pk, s} a_{p\ Pk, s})|\Omega, t\rangle$$

$$+e\sqrt{\frac{N}{V}}\sum_{\vec{p}s}\sum_{\vec{k}r}\frac{1}{\sqrt{2\omega_{\vec{k}}}}(\alpha_{\vec{k}r}\vec{\epsilon}_{\vec{k}r}e^{-i\omega_{\vec{k}}t} + \text{c.c.}) \cdot (a^\dagger_{p\ S, s} a_{p\ Pk, s}\langle S|\vec{j}_{em}|Pk\rangle + \text{h.c.})|\Omega, t\rangle$$

$$(4.87)$$

while for the Maxwell equation we may write:

$$i\dot{\alpha}_{\vec{k}r} - \frac{1}{2\omega_{\vec{k}}}\ddot{\alpha}_{\vec{k}r} = e\sqrt{\frac{N}{V}}\frac{\vec{\epsilon}^*_{\vec{k}r}}{\sqrt{2\omega_{\vec{k}}}}\frac{e^{i\omega_{\vec{k}}t}}{N}\sum_{ps}\langle\Omega, t|\{a^\dagger_{p\ S, s} a_{p\ Pk, s}\langle S|\vec{j}_{em}|Pk\rangle + \text{h.c.}\}|\Omega, t\rangle.$$

$$(4.88)$$

Using (4.86) we have:

$$i\frac{\partial}{\partial t}|\Omega, t\rangle = \sum_{\vec{p}s}\left[\frac{d}{dt}\cos\alpha_{\vec{p}} + \sum_k u_k \frac{d}{dt}\sin\alpha_k a^\dagger_{p\ Pk,s} a_{p\ Pk,s}\right]$$
$$\cdot \prod_{\vec{q}\neq\vec{p}s}\left[\cos\alpha_q + \sum_k u_k \sin\alpha_q a^\dagger_{qPk,s} a_{qS,s}\right]|0\rangle; \tag{4.89}$$

projecting (4.87) onto the perturbative vacuum (4.80) and noting that the matter amplitudes within the Fermi sphere do not depend on p we get:

$$i\frac{d}{dt}\cos\alpha_q = E_S\cos\alpha_q + e\sqrt{\frac{N}{V}}\sum_{\vec{k}r}\frac{1}{\sqrt{2\omega_{\vec{k}}}}(\alpha_{\vec{k}r}\vec{\epsilon}_{\vec{k}r}e^{-i\omega_{\vec{k}}t} + \text{c.c.})\cdot\sin\alpha_q u_k\langle S|\vec{j}_{em}|Pk\rangle, \tag{4.90a}$$

while projecting (4.87) onto the states $|k\rangle = \prod_{\vec{p},s}a^\dagger_{\vec{p}Pk,s}a_{\vec{p}S,s}|0\rangle$, we get:

$$i\frac{d}{dt}\sin\alpha_q = E_P\sin\alpha_q + e\sqrt{\frac{N}{V}}\sum_{\vec{k}r}\frac{1}{\sqrt{2\omega_{\vec{k}}}}(\alpha_{\vec{k}r}\vec{\epsilon}_{\vec{k}r}e^{-i\omega_{\vec{k}}t} + \text{c.c.})\cdot\cos\alpha_q u_k^*\langle Pk|\vec{j}_{em}|S\rangle. \tag{4.90b}$$

Finally substituting (4.86) in (4.88) we readily obtain:

$$i\dot{\alpha}_{\vec{k}r} - \frac{1}{2\omega_{\vec{k}}}\ddot{\alpha}_{\vec{k}r} = e\sqrt{\frac{N}{V}}\frac{\vec{\epsilon}^*_{\vec{k}r}}{\sqrt{2\omega_{\vec{k}}}}e^{i\omega_{\vec{k}}t}\frac{1}{N}\sum_{ps}[u_k\sin\alpha_{\vec{p}}\cos\alpha_{\vec{p}}\langle S|\vec{j}_{em}|Pk\rangle + \text{c.c.}]. \tag{4.90c}$$

The remarkable, but expected feature of the differential system (4.90) is its independence on the momentum variable $\vec{p}$. Naturally this system does not hold for **all** momenta, but only for momenta up to the Fermi momentum p_F, defined by the normalization equation:

$$\langle\Omega, t|N|\Omega, t\rangle\langle\Omega, t|\sum_{ps}(a^\dagger_{p\ S,s}a_{p\ S,s} + a^\dagger_{p\ Pk,s}a_{p\ Pk,s})|\Omega, t\rangle =$$
$$= 2\sum_{\vec{p}\leq\vec{p}_F}(\cos^2\alpha_{\vec{p}} + \sin^2\alpha_{\vec{p}}) = \text{N}, \tag{4.91}$$

in agreement with (4.83). The noted momentum independence allows us to set $\alpha_{\vec{p}} = \alpha$ for $|\vec{p}| < p_F$ and the system (4.90) takes up exactly the form (4.11), that we have already analysed for ^{4}He, provided one makes the identification $a_0 = \cos\alpha$ and $a_{nk} = u_k\sin\alpha$. Following the steps that took us from (4.11) all the way to (4.37) we learn that the ^{3}He liquid has exactly the same electrodynamical coherence properties as ^{4}He. In particular for each atoms it develops a gap $\delta \simeq 1\ K$ [see (4.37)]. Why then its properties are so different from those of ^{4}He? We can easily see that such a gap only shifts the single particle spectrum of ^{3}He by the quantity δ, however this shift is completely inconsequential for, as far as the fermions (the ^{3}He atoms) are concerned, the structure of the ground state does remain that of the free system with an apparently irrelevant difference: its single particle wave-function contains a small but finite P-wave admixture and the ^{3}He is no more a compact little ball. All this means that ^{3}He still remains a typical Fermi-liquid with a particle spectrum that is strongly affected by short-range forces. In fact, if one represents the single quasi-particle energy as $\epsilon_{\vec{k}} = \frac{k^2}{2m^*}$, in the vicinity of the Fermi surface ($|k| \simeq k_F$) experimentally one finds $m^* \simeq 3\, m_3$, a huge mass renormalization!

With regard to superfluidity, one realizes that what makes a fermionic system a superfluid is dramatically different from what is responsible for the superfluidity of a Bose liquid. In the latter case, as we have seen, one needs the condensation of a coherent classical field, in the former it is necessary that the spectrum of low energy excitations, which in view of the Pauli principle must lie around the Fermi surface, acquires an energy gap Δ that represents the minimum energy needed for an external agent to perturb the system. If such an external agent is a thermal fluctuation, one expects superfluidity to be lost when the typical energy of the fluctuation ($\simeq k\, T$) equals Δ.

Having understood why the e.m. coherence, which has such an outstanding effect on ^{4}He, leaves ^{3}He practically unaffected, we are still left with the big puzzle of its superfluidity at very low temperature. The mystery becomes thicker when we learn from experiment that the superfluidity gap shows important anisotropies, which in the technical jargon are represented by saying that ^{3}He exhibits a "P-wave superfluidity". But how does the Ockham's razor fare with this strange phenomenon that an almost perfect ball like ^{3}He finds a way to interact with its environment made of equally perfect balls in an anisotropic way? How is it possible that the nuclear spins, whose electromagnetic coupling is miserably small, succeed in communicating at large distances and produce a gap?

Postponing the detailed mathematical analysis of ^{3}He superfluidity to Chapter 6, where the closely related problem of superconductivity will also be analysed, I wish to indicate an important clue in this direction arising from the weak electrodynamical coherence of ^{3}He that we have discussed in this Section.

The mentioned small P-wave admixture in the wave-functions of the electrons in the CGS suggests that the system may develop a non-negligible electronic angular momentum. In fact one has:

$$\vec{L} = N \sum_{khn} a_{kn}^* \langle nPk|\vec{L}|nPh\rangle a_{hn}, \qquad (4.92)$$

which follows from the definition (4.10) of the wave-field. Using the Wigner-Eckart theorem:

$$\langle nPk|L_i|nPh\rangle = i\epsilon_{kih} \qquad (4.93)$$

and the expression (4.24) valid in the CGS, we compute:

$$\langle L_i\rangle = \theta^2 i\epsilon_{kih} u_k^* u_h, \qquad (4.94)$$

which on the 4-dimensional manifold spanned by u_k takes up all directions and all magnitudes from 0 the maximum value $N\theta^2$.

Normally $\langle \vec{L}\rangle$ vanishes, for otherwise the associated magnetic field:

$$\vec{B}_e = \frac{e}{2m_e}\langle \vec{L}\rangle \frac{N}{V} \qquad (4.95)$$

would increase the energy density of the state by the magnetic energy density $\frac{1}{2}\vec{B}_e^2$. However, if due to the existence of some other interactions the system finds it advantageous to develop a non vanishing $\langle \vec{L}\rangle$, it will do so without any difficulty, for the quantum mechanics of an open system will always tend to the true ground state. This happens for instance when an external magnetic field is switched on, in this case $\vec{B}_e$ will decrease the overall magnetic energy by aligning in a direction opposite to $\vec{H}$. For ^{3}He, even in the absence of an external magnetic field, the short-range hyperfine interaction:

$$H_{hf} = g\vec{L}_e \cdot \vec{S}_{^3He} \qquad (4.96)$$

couples the electronic angular momentum to the spin of the nucleus. In this way energy is minimized by having all nuclei (within a CD) point their spins opposite to the common direction of the electrons' angular momentum $\langle \vec{L}_e\rangle$ and as a result a long range pairing interaction for nuclear spins emerges that is attractive when the spins are parallel. This looks precisely like what is needed for accounting for the strange anisotropies of ^{3}He superfluidity. But we will have to analyse the problem from scratch, after having developed the elegant kinematical framework due to Bardeen, Cooper and Schrieffer (BCS). This will be done in Chapter 6.

Exercises of Chapter 4

$\langle \mathbf{4.1}\rangle$: Derive (4.4).

$\langle 4.2 \rangle$: Derive the numerical estimate (4.39).

$\langle 4.3 \rangle$: Derive (4.56).

$\langle 4.4 \rangle$: Do the calculations leading to (4.67) explicitly.

References to Chapter 4

1. R. Mele, G. Preparata and S. Villa, preprint MITH 94/6.
2. L.D. Landau, *J.Phys. USSR* **5** (1941) 71;
   and ibid. **11** (1947) 91.
3. for a discussion see E. C. Svensson, in in *Elementary Excitations in Quantum Fluids*, eds. K. Ohbayashi and M. Watabe (Springer, Berlin, 1989).
4. P.V. Mc Clintock, *Nature* **347** (1990) 233.
5. E. Del Giudice, M. Giuffrida, R. Mele and G. Preparata, *Europhys. Lett.* **14** (1991) 463.
   E. Del Giudice, M. Giuffrida, R. Mele and G. Preparata, *Phys. Rev.* **B43** (1991) 5381.
6. F. Mezei, *Phys. Rev. Lett.* **44** (1980) 1601.
7. A.D.B. Woods and E.C. Svensson *Phys. Rev. Lett.* **41** (1978) 974.
8. J. Maynard *Phys. Rev.* **B14** (1976) 3868.
9. L.D. Landau and E. M. Lifšits, *Course of Theoretical Physics Vol.5 - Statistical Physics* (Pergamon Press, Oxford, 1969).
10. L. Onsager, *Il Nuovo Cimento* **6** (1949) 246.
11. E. Del Giudice, C.P. Enz, R. Mele and G. Preparata, *Il Nuovo Cimento* **15 D** (1993) 3613.
12. F. London, *Superfluids* (Wiley, New York, 1954).
13. P. A. Whitlock, D. M. Ceperley, G. V. Chester and M. H. Kalos, *Phys. Rev.* **B 19** (1979) 5598.

Chapter 5

QFT OF PLASMAS: IDEAL AND REAL

5.1 Models of Quantum Plasmas (QP)

Plasmas – the fourth state of matter – are ubiquitous in nature. They occur every time the constituents of a matter system are charged. The global neutrality of stable matter requires that any plasma be composed of an equal number of positive and negative particles. The definition, that has just been given, must however be sharpened for, if one does not specify the space distances and the time intervals in which the dynamics of the matter system is being followed through, all matter is ultimately a plasma, being composed of charged elementary constituents, electrons and nuclei, if only we wish to leave out quarks and gluons. Thus when speaking about a plasma we must always be clear as to the space-time domains over which its opposite charges perform **independent** oscillations, thus lifting the overall neutrality of correlated opposite charges to the electromagnetic interaction.

The simplest type of plasma, the ideal plasma, consists of a set of N charged particles of charge eQ ($Q = -1$ for the electron), enclosed in a volume V, moving in a homogeneous charged fluid, whose total charge is $-NeQ$, as demanded by overall neutrality. The classical equations of motion for the N charged particles are

$$m\ddot{\vec{x}}_k = Q \, e \, \vec{E}\left(\vec{x}_k\right) \qquad (k = 1, ...N) \qquad (5.1)$$

$\vec{E}(\vec{x})$ is the electrostatic field resulting from the given plasma configuration. At equilibrium $\vec{E}(\vec{x}_k) = 0$, but if we perturb the positions of the k-th charge by the small displacement $\vec{\xi}_k$ we have from (5.1)

$$\ddot{\vec{\xi}}_k = \frac{Qe}{m} \, \vec{E}_{\xi}(\vec{x}_k + \vec{\xi}_k). \qquad (5.2)$$

93

For small $\vec{x}$, for the non-zero electric field $\vec{E}_{\vec{\xi}}(\vec{x}_k + \vec{x})$ caused by the displacement $\vec{\xi}$ we may write:

$$\vec{E}_{\vec{\xi}}(\vec{x}_k + \vec{x}) = A \,\hat{\xi}\,(\hat{\xi} \cdot \vec{x}) + O(\vec{x}^2), \tag{5.3}$$

$\hat{\xi}$ being the unit vector in the direction of $\vec{\xi}$. It is now very easy to obtain the value of A: from the Poisson's equation we have in fact

$$\nabla \cdot \vec{E}(\vec{x}) = \rho_{fluid} + \rho_{part} = -Q\,e\,\frac{N}{V}\bigg|_{\vec{x} \simeq \vec{x}_k}. \tag{5.4}$$

On the other hand from (5.3) and (5.4) we obtain

$$\nabla \cdot \vec{E}_{\vec{\xi}}\,(\vec{x}) = A = -Q\,e\,\frac{N}{V}, \tag{5.5}$$

and noting that from (5.3)

$$\vec{E}_{\vec{\xi}_k} = -Q\,e\,\frac{N}{V}\,\vec{\xi}_k, \tag{5.6}$$

(5.2) becomes

$$\ddot{\vec{\xi}}_k = -\frac{N}{V}\,\frac{Q^2 e^2}{m}\,\vec{\xi}_k, \tag{5.7}$$

showing that a charged particle, once perturbed from its equilibrium position, oscillates around it with a well defined frequency, the plasma frequency

$$\omega_P = \frac{Qe}{\sqrt{m}}\,\left(\frac{N}{V}\right)^{\frac{1}{2}}. \tag{5.8}$$

This is all there is in the dynamics of the small oscillations of charged particles around their classical equilibrium position: each particle behaves as a 3-dimensional harmonic oscillator with frequency ω_P. The classical Hamiltonian for each charged particle can now be written down in a straightforward manner:

$$H_{plasma} = \frac{\vec{p}^2}{2m} + \frac{1}{2}\,m\,\omega_P^2\,\vec{\xi}^2 \tag{5.9}$$

where $\vec{p} = m\,\dot{\vec{\xi}}$.

The QFT of the ideal plasma can be easily constructed by introducing first the matter wave-field

$$\psi(\vec{x}, \vec{\xi}; t) = \sum_{\vec{n}} \psi(\vec{x}, t)_{\vec{n}}\,\langle \vec{\xi} | \vec{n} \rangle, \tag{5.10}$$

where $\vec{x}$ spans the set of equilibrium position, and $\langle \vec{\xi}|\vec{n}\rangle$ are the space eigenfunctions of the 3-dimensional harmonic oscillator belonging to the eigenvalues $\vec{n} = (n_1, n_2, n_3)$ (n_i are non-negative integers). The commutation relations read as usual

$$[\psi^+(\vec{x},t)_{\vec{n}}, \psi(\vec{y},t)_{\vec{n}'}] = \delta_{\vec{n}\vec{n}'}\delta^3(\vec{x}-\vec{y}). \tag{5.11}$$

The Hamiltonian for the interaction at the quantized electromagnetic field, according to Eq. (1.72), has the form:

$$\begin{aligned} H_{int}^{(1)} &= e\int_{\vec{x}\vec{\xi}}\psi^+(\vec{x},\vec{\xi};t)\vec{J}\psi(\vec{x},\vec{\xi};t)\vec{A}(\vec{x},t) \\ &= e\int_{\vec{x}}\sum_{\vec{n}\vec{n}'}\langle\vec{n}|\vec{J}|\vec{n}'\rangle\psi^+(\vec{x},t)_{\vec{n}}\psi(\vec{x},t)_{\vec{n}'}\vec{A}(\vec{x},t) \end{aligned} \tag{5.12}$$

where

$$\langle\vec{n}|\vec{J}|\vec{n}'\rangle = Q\langle\vec{n}|\frac{\vec{p}}{m}|\vec{n}'\rangle, \tag{5.13}$$

as follows from the minimal shift performed on the Hamiltonian (5.9). For future convenience we record here the relations between the oscillator operators $\vec{a}$ and $\vec{a}^+$, with commutation relations

$$[a_i, a_j^+] = \delta_{ij}, \tag{5.14}$$

and the operator $\vec{\xi}$ and $\vec{p}$. One has

$$\vec{a} = (\frac{m\omega_P}{2})^{\frac{1}{2}}\vec{\xi} + i(\frac{1}{2m\omega_P})^{\frac{1}{2}}\vec{p}, \tag{5.15a}$$

$$\vec{a}^+ = (\frac{m\omega_P}{2})^{\frac{1}{2}}\vec{\xi} - i(\frac{1}{2m\omega_P})^{\frac{1}{2}}\vec{p}, \tag{5.15b}$$

or, by inverting (5.15),

$$\vec{\xi} = (\frac{1}{2m\omega_P})^{\frac{1}{2}}(\vec{a} + \vec{a}^+), \tag{5.16a}$$

$$\vec{p} = i(\frac{m\omega_P}{2})^{\frac{1}{2}}(\vec{a}^+ - \vec{a}). \tag{5.16b}$$

Note that the second-order term $H^{(2)}$ can be neglected for, referring to Eq. (3.46), the intermediate states that contribute most to it are those which, in order to avoid double counting, should be excluded from the intermediate states' sum.

The space-independent CE's, following from the Hamiltonian:

$$H = \sum_{\vec{n}} \int_{\vec{x}} \psi^+(\vec{x},t)_{\vec{n}} \psi(\vec{x},t)_{\vec{n}} \omega_P \sum_k \left(\vec{n}_k + \frac{1}{2}\right) + H_{int} + H_{em} \qquad (5.17)$$

can be readily written down:

$$i\frac{\partial \psi}{\partial t}(t)_{\vec{n}} = \omega_P \sum_k \left(\vec{n}_k + \frac{1}{2}\right)\psi(t)_{\vec{n}} + eQ\left(\frac{N}{V}\right)^{\frac{1}{2}} \sum_{\vec{k}r\vec{n}'} \frac{1}{(2\omega_{\vec{k}})^{\frac{1}{2}}} \qquad (5.18a)$$

$$\cdot [\alpha_{\vec{k}r}\vec{\epsilon}_{\vec{k}r}e^{-i\omega_{\vec{k}}t} + h.c.]\langle \vec{n}|\frac{\vec{p}}{m}|\vec{n}'\rangle \psi(t)_{\vec{n}'},$$

$$-\frac{1}{2\omega_{\vec{k}}}\ddot{\alpha}_{\vec{k}r}(t) + i\dot{\alpha}_{\vec{k}r}(t) - \frac{Q^2 e^2}{\omega_{\vec{k}}}\left(\frac{N}{V}\right)\lambda\alpha_{\vec{k}r}$$

$$= \frac{Qe}{(2\omega_{\vec{k}})^{\frac{1}{2}}}\left(\frac{N}{V}\right)^{\frac{1}{2}}\vec{\epsilon}_{\vec{k}r}^*\, e^{i\omega_{\vec{k}}t}\,. \qquad (5.18b)$$

$$\cdot \sum_{\vec{n}\vec{n}'}\langle \vec{n}|\frac{\vec{p}}{m}|\vec{n}'\rangle \psi(t)_{\vec{n}}^* \psi(t)_{\vec{n}'}$$

By going to the interaction representation

$$\psi(t)_{\vec{n}} = \phi(t)_{\vec{n}}e^{-i\omega_P \sum_k (n_k + \frac{1}{2})t}, \qquad (5.19)$$

and noting that from (5.16b) we can write

$$\langle \vec{n}|\frac{\vec{p}}{m}|\vec{n}'\rangle = i\left(\frac{\omega_P}{2m}\right)^{\frac{1}{2}}\langle \vec{n}|(\vec{a}^+ - \vec{a})|\vec{n}'\rangle, \qquad (5.20)$$

we immediately realize that in the "rotating wave approximation" only those modes of the e.m. field need be considered for which

$$\omega_{\vec{k}} = \omega_P = \frac{Qe}{(m)^{\frac{1}{2}}}\left(\frac{N}{V}\right)^{\frac{1}{2}}. \qquad (5.21)$$

Introducing the dimensionless time $\tau = \omega_P t$, we rewrite the system (5.18) as

$$\dot{\phi}_{\vec{n}}(\tau) = \frac{1}{2}\sum_{|\vec{k}|=\omega_P}\sum_{\vec{n}'r}\langle \vec{n}|\alpha_{\vec{k}r}\vec{\epsilon}_{\vec{k}r}\vec{a}^+ - \alpha_{\vec{k}r}^*\vec{\epsilon}_{\vec{k}r}\vec{a}|\vec{n}'\rangle\phi_{\vec{n}'}(\tau) \qquad (5.22a)$$

$$\frac{1}{2}\ddot{\alpha}_{\vec{k}r} - i\dot{\alpha}_{\vec{k}r} + m\lambda\alpha_{\vec{k}r} = \frac{i}{2}\vec{\epsilon}_{\vec{k}r}^* \sum_{\vec{n}\vec{n}'}\langle \vec{n}|\vec{a}|\vec{n}'\rangle\phi_{\vec{n}}^*(\tau)\phi_{\vec{n}'}(\tau), \qquad (5.22b)$$

which by defining the state

$$|\phi\rangle = \sum_{\vec{n}} \phi_{\vec{n}}(\tau)|\vec{n}\rangle, \tag{5.23a}$$

and the e.m. amplitude

$$\vec{A} = \left(\frac{3}{8\pi}\right)^{\frac{1}{2}} \sum_r \int d\Omega_k \alpha_{\vec{k}r} \vec{\epsilon}_{\vec{k}r}, \tag{5.23b}$$

can be given the more manageable form:

$$\frac{\partial}{\partial t}|\phi\rangle = \left(\frac{2\pi}{3}\right)^{\frac{1}{2}} (\vec{A}\vec{a}^+ - \vec{A}^*\vec{a})|\phi\rangle, \tag{5.24a}$$

$$\dot{\vec{A}} + \frac{i}{2}\ddot{\vec{A}} + im\lambda\vec{A} = -\left(\frac{2\pi}{3}\right)^{\frac{1}{2}}\langle\phi|\vec{a}|\phi\rangle. \tag{5.24b}$$

Recalling that the short-time dynamics takes off from $|\phi(0)\rangle = |0\rangle$, differentiating (5.24b) and using (5.24a) with $|\phi\rangle = |0\rangle$, we easily obtain (as argued above, we may neglect the photon mass term $m\lambda$)

$$\ddot{A}_j + \frac{i}{2}\dddot{A}_j = -\left(\frac{2\pi}{3}\right)\langle 0|\ [a_j, a_l^+]\ |0\rangle \quad A_l = -\left(\frac{2\pi}{3}\right)A_j \tag{5.25}$$

which is precisely in the canonical form (3.17) with $\mu = 0$ and $g^2 = \left(\frac{2\pi}{3}\right)$. The critical coupling constant for our plasma is (see (3.19)):

$$g_c^2 = \frac{16}{27} = .593 < g^2 = \left(\frac{2\pi}{3}\right) = 2.1, \tag{5.26}$$

the ideal plasma "runs away"! Whereto?

Defining

$$\alpha_k = \langle\phi|a_k|\phi\rangle \tag{5.27}$$

the system (5.24) becomes $\left(g_0 = \left(\frac{2\pi}{3}\right)^{\frac{1}{2}}\right)$

$$\dot{\alpha}_k = g_0 A_k, \tag{5.28a}$$

$$\dot{A}_k + \frac{i}{2}\ddot{A}_k = -g_0\alpha_k, \tag{5.28b}$$

admitting the conserved quantity

$$Q = \sum_k \left\{ A_k^* A_k + \frac{i}{2}(A_k^* \dot{A}_k - \dot{A}_k^* A_k) + \alpha_k^* \alpha_k \right\}, \tag{5.29a}$$

while the Hamiltonian is easily computed to be:

$$\frac{E}{N\omega_P} = H = Q + \sum_k \left[\frac{1}{2} \dot{A}_k^* \dot{A}_k - ig(A_k^* \alpha_k - A_k \alpha_k^*) \right]. \tag{5.29b}$$

In order to see whether there are "extremal" stationary solutions, we set as usual

$$\alpha_k = \alpha u_k e^{i\psi}, \tag{5.30a}$$

$$A_k = A u_k e^{i\phi}, \tag{5.30b}$$

with α and A positive constants and u_k a complex vector.

Substituting (5.30) in Eqs. (5.28) one obtains

$$\phi - \psi = \frac{\pi}{2}, \tag{5.31a}$$

$$\alpha = g_0 \frac{A}{\dot{\phi}}, \tag{5.31b}$$

$$\frac{1}{2}\dot{\phi}^3 - \dot{\phi}^2 + g_0^2 = 0, \tag{5.31c}$$

and from the condition $Q = 0$, as implied by the assumed start from the PGS, we get

$$1 - \dot{\phi} + \frac{g_0^2}{\dot{\phi}^2} = 0. \tag{5.32}$$

It is easy to check (Ex. $\langle 5.1 \rangle$) that for $g_0^2 = \left(\frac{2\pi}{3} \right)$ it is impossible to satisfy both (5.31c) and (5.32).

This strange result simply means that the energy of the ideal Quantum Plasma (QP) does not have a minimum, or, put differently, that the ideal QP does not exist as a physical system. This latter fact, however, is not unexpected, for the arbitrarily low energy states that can be reached by the dynamics expressed by (5.28) are associated with arbitrarily large amplitudes A and α, or arbitrarily large plasma oscillation amplitudes, as one learns from the relations (5.16) which imply:

$$\langle \vec{\xi}^2 \rangle = \frac{1}{m\omega_P}\,\alpha^2. \tag{5.33}$$

But our derivation of the plasma Hamiltonian (5.9) indicates clearly that, even without questioning the rather strong assumption of the homogeneous fluid, our approximation must break down when

$$\langle \vec{\xi}^2 \rangle^{\frac{1}{2}}_{max} \simeq a = \left(\frac{V}{N}\right)^{\frac{1}{3}}, \tag{5.34}$$

i.e. when the plasma oscillations become of the same order as the average interparticle distance a.

In order to make models for more realistic plasmas, let us compute the breakdown amplitude α_{max}, obtained by combining (5.33) and (5.34) for the two interesting cases of an electron plasma, and a typical plasma of nuclei of effective charge $Z_{eff} \simeq 1$ and mass $m_N \simeq 50\,\text{GeV}$.

By using the definition (5.8) of ω_P one derives for α_{max} the relation

$$\alpha_{max} = (m\omega_P)^{\frac{1}{2}}\left(\frac{V}{N}\right)^{\frac{1}{3}} = (ma)^{\frac{1}{4}} e^{\frac{1}{2}}. \tag{5.35}$$

Taking $a \simeq 2.5\,\mathring{A}$, for electrons one computes

$$\alpha^e_{max} \simeq 2.7, \tag{5.36a}$$

and for typical nuclei

$$\alpha^N_{max} \simeq 50. \tag{5.36b}$$

These simple calculations show very clearly the direction in which one should go in order to turn the ideal QP into a realistic physical system. For electron plasmas, according to (5.36a), the plasma oscillations remain of rather small amplitude, thus a model of the type discussed in Section (4.3), in which the coherent dynamics involves only the ground state and the first excited states ($|\vec{n}| = 1$) leaving out all other excited states ($|\vec{n}| > 1$), can be a rather adequate approximation, especially if the coupling constant g is realistically modified from its ideal plasma value $g_0 = \left(\frac{2\pi}{3}\right)^{\frac{1}{2}}$. In order to understand this latter point a little better, let us note that one of the consequences of going beyond the homogeneous fluid approximation is to change the frequency of small-amplitude oscillations from the ideal value ω_P to some other value ω_R (the "real value"). In general one will have $\omega_R > \omega_P$, due to the fluid dishomogeneities that will generally enhance the fluid charge density in the vicinity of the plasma equilibrium positions. In fact this is exactly what happens when the plasma goes towards an "atomic configuration", in which the electrons oscillate around their own atomic nucleus. If this is the case, it is sufficient to go through the steps that led from (5.18) to (5.24) to find out that

$$g^2 = \left(\frac{2\pi}{3}\right)\left(\frac{\omega_P}{\omega_R}\right)^2, \tag{5.37}$$

implying that these more realistic plasmas will "run away" when:

$$\left(\frac{\omega_R}{\omega_P}\right) < \sqrt{\frac{2\pi}{3}\frac{27}{16}} = 1.88, \tag{5.38}$$

and this leads us to expect many real plasmas of CM to possess coherent dynamical configurations.

For plasmas with large oscillations amplitudes, like the typical plasma of nuclei, a viable model can be constructed by adding to the plasma Hamiltonian a term:

$$H_c = \omega_R \left(\sum_k \frac{a_k^+ a_k}{\alpha_{max}^2}\right)^n \tag{5.39}$$

where n is a large integer that we may send to ∞ at the end. The effect of H_c is evidently to strongly suppress coherent oscillation amplitudes $\alpha^2 \geq \alpha_{max}^2$. The addition of (5.39) modifies only Eq. (5.28a) of the ideal QP system, which now becomes

$$\dot{\alpha}_k = gA_k - in\left(\sum_j \frac{\alpha_j^* \alpha_j}{\alpha_{max}^2}\right)^{n-1} \frac{\alpha_k}{\alpha_{max}^2} \tag{5.40}$$

leaving (5.28b) and (5.29) unchanged. Making now the Ansatz (5.30) we obtain

$$\phi - \psi = \frac{\pi}{2}, \tag{5.41a}$$

$$\dot{\psi} = g\frac{A}{\alpha} - \frac{n}{\alpha_{max}^2}\left[\frac{\alpha^2}{\alpha_{max}^2}\right]^{n-1}, \tag{5.41b}$$

$$\frac{1}{2}\dot{\phi}^2 - \dot{\phi} + g\frac{\alpha}{A} = 0, \tag{5.41c}$$

and from the conservation equation (5.29a) we get

$$A^2(1 - \dot{\phi}) + \alpha^2 = 0. \tag{5.42}$$

The solution of (5.41) with the constraint (5.42) is quite simple. Setting $x = \frac{\alpha}{A}$ (5.41c) implies

$$\dot{\phi} = 1 + \sqrt{1 - 2gx}, \tag{5.43a}$$

while (5.42) yields

$$x^2 = \sqrt{1 - 2gx}. \tag{5.43b}$$

From $\dot{\phi} = \dot{\psi}$ (following from (5.41a)) one gets

$$A = \left(\frac{\alpha_{max}}{x}\right)\left[\frac{\alpha_{max}^2}{n}\left(\frac{g}{x} - \dot{\phi}\right)\right]^{\frac{1}{2(n-1)}} \rightarrow_{n\to\infty} \left(\frac{\alpha_{max}}{x}\right). \tag{5.43c}$$

For the energy of this state we obtain from (5.29b)

$$\frac{E}{N\omega_R} = \alpha_{max}^2 \frac{[1 + x^2 - 3gx]}{x^2}, \tag{5.44}$$

whose behaviour as a function of g is plotted in Fig. 5.1.

As expected, for $g^2 > g_c^2$ the energy is negative: this state is the real GS. It is gratifying to note that the physics that emerges is quite transparent: the ground state of the charged particles is a coherent state in which the particles oscillate in phase with the e.m. field of classical (complex) amplitude

$$\alpha_k = u_k \alpha_{max} e^{-i\omega_R(1 - \dot{\phi})t}, \tag{5.45}$$

whose maximum value is just α_{max} and whose frequency gets renormalized from the "free" value ω_R as

$$\omega_r = \omega_R(1 - \dot{\phi}) < \omega_R, \tag{5.46}$$

leading to the by now familiar phenomenon of the trapping of the e.m. field inside the plasma.

All realistic models of QP's will somehow lie between the two extremes we have just discussed. However from the preceding analysis one thing can be stated with definiteness: the nonsensical dynamics of the ideal QP does get completely removed by a few simple, reasonable and realistic considerations, that lead us to some rather

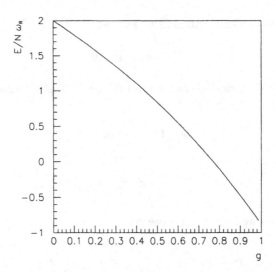

Fig. 5.1. The energy gain per particle (Eq. (5.44)) for a "realistic" large amplitude plasma as a function of the coupling g.

well defined models whose physics not only is neither pathological nor contrived but completely conforms to the general ideas of coherence discussed and analysed in the first Chapters.

5.2   Electrons' plasmas in metals: the gas-liquid transition and the Ohm's law

Even if one has become used to the peculiar behaviour of electrons in metals, I believe that there should be a general agreement that we still **do** miss the "raison d' être" of the existence of a metal and of the strange properties of its conduction electrons. To the CM physicist that will be certainly surprised, if not utterly offended by this statement I would like to emphasize that I recognize that we **do** possess a sophisticated and highly successful framework to deal with very subtle problems of the dynamics of the electrons, but to my mind these theoretical tools have never amounted to more than an extremely cunning "kinematics", whose foundation on the undisputed bases of a complete QED theory has **never** been either proved or quali- tatively understood. And the reason for this, as I shall argue below, is to be sought in the inability of GACMP to account for quantum electrodynamical coherence.

In this Section we shall deal with the conduction electrons of a simple metal, while an analysis of crystal structure and its relation to the coherent motion of nuclei will appear in Section (5.3).

For simplicity we shall consider a model with a single valence electron. The lattice structure of the metal will be completely neglected, for it is not expected to play an important rôle on the physics of the electrons' conduction, as electrons conduce electricity also in the liquid state.

Our electrons will then form a Fermi-liquid, structurally similar (but for the very different mass) to the ^{3}He-system considered in Section (4.4). The wave-field belonging to these electrons can then be written (χ_s is the usual Pauli spinor)

$$\psi(\vec{x}, \vec{\xi}; t) = \frac{1}{\sqrt{V}} \sum_{\vec{p} s \vec{n}} a_{\vec{p} s \vec{n}} e^{i \vec{p} \cdot \vec{x}} \chi_s \left\langle \vec{\xi} \middle| \vec{n} \right\rangle, \tag{5.47}$$

with the anticommutation relations (s is the dichotomic spin variable)

$$\{a_{\vec{p} s \vec{n}}, a_{\vec{p}' s' \vec{n}'}^+\} = \delta_{\vec{p} \vec{p}'} \delta_{ss'} \delta_{\vec{n} \vec{n}'}. \tag{5.48}$$

According to the above discussion, in the $\vec{n}$-sum we shall retain only the terms $\vec{n} = 0$ and $|\vec{n}| = 1$. Setting the density $\frac{N}{V} = \left(\frac{1}{a}\right)^3$ (typically $a \simeq 2.5\text{Å}$, corresponding to $\frac{N}{V} \simeq 6.4 \times 10^{22} \text{cm}^{-3}$), the PGS of our system will be of the type (4.81), where according to (4.83) the Fermi momentum in the continuum approximation $\left(\sum_{\vec{p}} \to \frac{V}{(2\pi)^3} \int d^3p\right)$ is given by

$$p_F = \frac{(3\pi^2)^{\frac{1}{3}}}{a}. \tag{5.49}$$

Note that the maximum momentum is $(2)^{\frac{1}{3}} p_F$: the conduction electron "band" is, as we know well, half filled. Assuming the free particle energy spectrum

$$\epsilon_{\vec{p}} = \frac{\vec{p}^2}{2m_e}, \tag{5.50}$$

the Fermi energy is

$$E_F = \frac{p_F^2}{2m_e} \simeq 5.6 \, \text{eV}. \tag{5.51}$$

The excited band, corresponding to the states $|\vec{n}| = 1$, comprises three similar structures (one for each of the three different configurations with $|\vec{n}| = 1$) displaced above by the energy $\omega_R > \omega_P$, where

$$\omega_P = \frac{e}{(m_e)^{\frac{1}{2}}} \left(\frac{1}{a}\right)^{\frac{3}{2}} = 9.6 \, \text{eV}, \tag{5.52}$$

implying a good band separation.

Let us now try to estimate $\frac{\omega_R}{\omega_P}$ which, according to (5.37), plays a crucial rôle in the coherence properties of the electron plasma. In order to achieve this we proceed as follows: we approximate the positive charge distributions by a constant distribution over the atomic volume with radius $R_{atom} \simeq \frac{a}{2}$, and zero outside. Putting the "atomic" density in the expression (5.8) for the plasma frequency, we obtain

$$\frac{\omega_R}{\omega_P} = \left(\frac{a^3}{\frac{4\pi}{3} R_{atom}^3}\right)^{\frac{1}{2}} = 1.38 < 1.88, \tag{5.53}$$

which using (5.38) tells us that the electron plasma is above the threshold for "strong coherence", its coupling constant being

$$g = \left(\frac{2\pi}{3}\right)^{\frac{1}{2}} \frac{1}{1.38} = 1.05 > .77. \tag{5.54}$$

We need only go through the discussion of Section (4.4) to be reminded that the e.m. coherence properties for a fermion-plasma are exactly equal to those of a bosonic system, and the mathematical developments of Section (4.3) to determine that the gap induced by the e.m. coherent evolution is

$$\delta_{ep} \simeq 2\,\text{eV}, \tag{5.55}$$

as can be read from the graph in Fig. 4.6 setting $g \sim 1.05$ [Eq. (5.54)].

This essentially completes the mathematical analysis of the e.m. coherence properties of the conduction electron plasma of a simple metal. Let us now try to put the physics into a sharper focus.

The emergence of a gap δ_{ep} [Eq. (5.55)] at the densities in which the metal is condensed provides a rational answer to the important question why at appropriate temperatures the monoatomic metal gas finds it convenient to "shrink" to such densities, thereby causing the gas-liquid phase transition. It is clear that from the size of the gap it should be possible to predict the critical temperature; however in concrete cases this requires a detailed analysis of the statistical thermodynamics of the quantum fluctuations above the CGS which, as stressed time and again, is beyond the scope of this book. Here only a rough sketch of a possible strategy for determining the transition temperature T_C between liquid and gas will be presented.

For definiteness' sake we consider a monoatomic metal gas with mass $m \simeq 50$ Gev. We know from thermodynamics that the liquid-gas phase transition occurs when the Gibbs functions $H = F + pV$ ($F = E - TS$ is the free energy) of the system in the two different phases become equal, *i.e.* when

$$F_l + pV_l = F_g + pV_g. \tag{5.56}$$

Let's compute first the gas free energy assuming it to be perfect and to consist of point-like particles of mass m.

For F_g one has the well known expression:

$$F_g = -NT \, ln \left(\frac{ev_g}{N} \int \frac{d^3p}{(2\pi)^3} e^{-\frac{p^2}{2mT}} \right) = -NT \, ln \left(\frac{ev_g}{N} (\frac{mT}{2\pi})^{\frac{3}{2}} \right). \tag{5.57}$$

As for the liquid, from our discussion in Section (2.4) we recall that it consists of two fluids, the superfluid whose free energy is (its entropy is clearly zero)

$$F_s = E_s = -N_s \delta_{ep} \tag{5.58}$$

and the normal fluid consisting of $N_n = N - N_s$ particles which thermal fluctuations have removed from the condensate. From the point of view of thermodynamics, a rough picture paints this fluid as a disordered gaseous system at the density of the liquid. Thus from (5.57) we may write

$$F_n = -N_n T \, ln \left(\frac{ev_l}{N} (\frac{mT}{2\pi})^{\frac{3}{2}} \right). \tag{5.59}$$

Substituting in (5.56), and recalling that $V_g \gg V_l$, the equation that determines the transition temperature T_C reads:

$$-\rho_s \delta_{ep} = T \left[\rho_n \, ln \left(e(\frac{V}{N})_l (\frac{mT}{2\pi})^{\frac{3}{2}} \right) - ln \left(e(\frac{V}{N})_g (\frac{mT}{2\pi})^{\frac{3}{2}} \right) + 1 \right] \tag{5.60}$$

where we have introduced the fraction of superfluid $\rho_s = \frac{N_s}{N} = 1 - \rho_n$.

In order to (roughly) determine the fraction of normal fluid ρ_n we argue as follows: in the Coherence Domain (CD), that we approximate as a sphere of radius $\frac{\pi}{\omega_R} = R$, the gap $\delta_{ep}(r)$ is also a function of the distance r from the center of the CD, that varies like the intensity A^2 of the e.m. condensate. In the spherical approximation [see Section (3.5)] this latter behaves as $\left(\frac{\sin \frac{\pi r}{R}}{\frac{\pi r}{R}} \right)^2$, so that we may write

$$\delta_{ep}(r) = \delta_{ep}(0) \left(\frac{\sin \frac{\pi r}{R}}{\frac{\pi r}{R}} \right)^2. \tag{5.61}$$

Little effort is required to see that

$$\rho_n(T) \simeq \left[\frac{4\pi}{3} R^3\right]^{-1} 4\pi \int_0^R r^2 dr e^{-\frac{\delta(r)}{T}} \qquad (5.62)$$

correctly represents the T-dependent fraction of the normal incoherent fluid in the liquid, for the spherical CD can be thought of being composed of concentric shells inside which the ratio between the atoms in the normal and in the "super" fluid is given simply by the Boltzmann factor $e^{-\frac{\delta(r)}{T}}$. Indeed $\delta(r)$ is the energy difference between the CGS and the PGS, where the incoherent fluctuations live. We have now all is needed to compute T_C: putting numbers in [Ex. $\langle 5.2 \rangle$] (we set $p = 1$Atm) we obtain

$$T_C \simeq 800\text{K}, \qquad (5.63)$$

corresponding to a normal fluid fraction at the transition temperature $\rho_n(T_C) \simeq .5$.

In view of the roughness of the calculation, and of the neglect of possible other coherent processes that might involve other types of electrons (we shall see later that d-electrons' coherence may play an important rôle in ferromagnetism) (5.63) may certainly be looked upon as a remarkable success of this simple analysis. The finite value of $\rho_s(T_C)$ demonstrates that the transition is, as expected and observed, of the first order type.

To summarize this discussion, we have just seen explicitly that in a simple metal the gas-liquid transition is associated with the plasma of conduction electrons having accessed the CGS. This induces us to conjecture that **all** gas-liquid transitions at high temperatures are indeed related to the setting in of some kind or other of electrons' coherent processes. This view will find a corroboration in the analysis of water reported in Chapter 10.

In the CGS the plasma of conducting electrons behaves like a superfluid whose constituents are protected against populating the incoherent PGS (as independent gas-like quantum fluctuations) by the **average gap** δ_{ep}. One should in fact speak of an average gap, for the electron plasma arranges itself in a spatial lattice of Coherence Domains (CD's) of size

$$\lambda_{ep} = \frac{2\pi}{\omega_R} \simeq 10^{-5}\text{cm} = 1000\text{Å}. \qquad (5.64)$$

The incoherent, localized quantum fluctuations that the residual interactions with the environment (phonons, impurities etc.) excite out of the CGS thus cluster in the interstitial regions of the array of CD's, for there the gap is much smaller or vanishes completely. Note that in those interstitial areas the electrons tend to resume their "atomic" configurations.

Let us apply an external electric field $\vec{E}$ (that we orient along the x-axis) to our metal. The electrons well inside each CD will behave as the electrons of a perfect conductor, being unable to dissipate energy due to the protection bestowed upon them by the gap δ_{ep}. Now, from the macroscopic Maxwell equations it is easy to derive that well within a CD no non-vanishing $\vec{E}$ can be sustained at equilibrium, $\vec{E}$ can only be present, unscreened, in the interstitial regions where superfluidity is switched off and the electrons have resumed their "atomic" configurations. The existence of this web of interstitial areas causes the superfluid to slowly drift in the direction opposite to the external electric field $\vec{E}$, for in those domains the electrons will be accelerated by the unscreened electric field $\vec{E}$ and scattered by the "normal fluid" to reach an equilibrium velocity, v_D.

Such drift velocity v_D, to which there is associated a macroscopic electric current

$$j = \frac{Ne}{V}v_D \tag{5.65}$$

can be determined by simple energy considerations: let us call Δx the average distance between two successive collisions, the equilibrium condition can be obtained by equating the work done on the single electron by the external electric field with the average energy that the electron loses to the lattice through incoherent scattering. Thus we may write

$$eE\Delta x = \frac{1}{2}m_e v_D^2 \Gamma \Delta t = eEv_D\frac{\Delta t}{2}, \tag{5.66a}$$

or

$$v_D = \frac{eE}{m_e}\frac{1}{\Gamma}, \tag{5.66b}$$

where the space interval Δx between two collisions has been put equal to $\frac{v_D}{2}\Delta t$ (as it follows from constant acceleration between collisions), Γ denotes the rate of collisions of the single electron in the incoherent region and $\frac{1}{2}m_e v_D^2$ is the kinetic energy that the electron loses in each collision .

Comparing (5.65) and (5.66b) yields nothing else than the Ohm's law:

$$\vec{j} = \sigma\vec{E} \tag{5.67}$$

where the constant

$$\sigma = \frac{N}{V}\frac{e^2}{m_e}\frac{1}{\Gamma} = \frac{N}{V}\frac{e^2}{m_e}\tau \tag{5.68}$$

is the familiar conductivity, and $\tau = \frac{1}{\Gamma}$ is usually called the relaxation time.

From the standard analysis of scattering theory we may write

$$\frac{1}{\tau} = \frac{N}{V} v_F \epsilon(T) \sigma_{scatt} \tag{5.69}$$

where σ_{scatt} is the scattering cross section, $\frac{N}{V} v_F$ is the flux of scattering centers, $\epsilon(T)$ is the T-dependent average over the initial and the final states of the electrons (all lying around the Fermi surface) participating in the scattering process (see Ex. $\langle 5.3\rangle$) :

$$\epsilon(T) = \frac{2V}{N} \int_{E_F}^{\infty} dE \rho(E) \frac{e^{\frac{(E-E_F)}{T}}}{\left(e^{\frac{(E-E_F)}{T}} + 1\right)^2}. \tag{5.70}$$

Taking for σ_{scatt} the atoms' geometric cross section:

$$\sigma_{scatt} \equiv \pi R_{at}^2 = \frac{\pi}{4} a^2, \tag{5.71}$$

and computing (5.70) for a simple isotropic Fermi gas we finally get the very simple expression:

$$\tau = \frac{4}{\pi} \frac{a}{\epsilon(T) v_F} = \frac{4}{\pi^{\frac{1}{3}} 3^{\frac{2}{3}}} \left(\frac{1}{T}\right). \tag{5.72}$$

The $\frac{1}{T}$ dependence of (5.72), that gives rise to the well-known linear T-dependence of the resistivity $\rho = \frac{1}{\tau}$, is particularly pleasing. However, we may ask whether the actual value of τ we obtain in (5.72) has anything to do with physical observations. At room temperature ($T = 300$ K) we compute $\tau = 3.2 \cdot 10^{-14}$ sec, which should be compared with the relaxation times of simple metals such as Ag and Au: experimentally $\tau_{Ag} = 4.0 \cdot 10^{-14}$ sec and $\tau_{Au} = 3.0 \cdot 10^{-14}$ sec, close to our simple computation. Now, one could go about doing something interesting with the enormous phenomenology of electrical conduction of metals. But, once more, this is not the aim of this book and we will pass on to analyse other very interesting plasmas, the plasmas of ions in solids.

5.3   Plasmas of ions and crystal structure

Our search for quantum electrodynamical coherence effects in the plasmas of the valence electrons of simple metals has given strong indications that the basic features of conduction electrons, and indeed the very existence of the liquid metal phase, can all be explained by the peculiar structure of the solutions of the CE's, that one derives for the dynamics of the electron plasma wave-field interacting with the quantized e.m. radiation field. Our study has revealed, so to speak, the germs of a general explanation of the existence and basic properties of high temperature liquids, such as metals, in the coherence properties of the global quantum mechanical wave-functions, that describe a collective field dynamics of the plasma charges oscillating in phase with a "trapped" classical configuration of the e.m. field.

We are now going to investigate whether the fundamental spatial order of the solid state – the crystal structure – may find its explanation in the emergence of electrodynamical coherence for the plasmas of ions that, though assumed to make up the charged fluid equilibrating the electrons' plasmas, are usually neglected. In Section (4.3) – in the case of ^{4}He – we have already had a glimpse of the growth of a solid from the liquid phase. There we have seen that a necessary condition for "hooking" the nuclear notion to the e.m. radiative field is that physical situations arise in which the Born-Oppenheimer approximation breaks down. In fact, it is only when the oscillations of the nuclei cannot be completely followed by the charge compensating electrons that the nuclei start emitting and absorbing radiation of their own frequency. And if the interaction and density is "strong enough", below a given temperature – the critical temperature – they will be able to access a coherent, collective state of lower energy, hence stable.

For ^{4}He we have identified the failure of the Born-Oppenheimer approximation in the deformation of the electronic shell that is caused by the too close approach of the oscillating atoms. What about a typical metal? From what we have learned in the preceding Section there is one fundamental difference between ^{4}He and a liquid metal besides the different temperatures at which they exist in the liquid phase. In the former case the elementary constituents, the ^{4}He-atoms, are neutral atoms in a slightly excited state, in the latter the oscillating nuclei are effectively in a ionized state, the valence electrons being delocalized in the "conduction band", whose large energy spread is just another piece of evidence that they belong to the global system and not to the single atoms, or to small clusters of atoms. Thus the global "covalence" of the conduction electrons, a consequence of their coherent dynamics, violates the Born-Oppenheimer approximation for the simple reason that the conduction electrons can only partly follow the oscillations of the ions, being primarily engaged in their own oscillations. This latter point, however, hides a subtlety: in order to be really valid it is necessary that the oscillations of the ions be strictly confined in space, where they may be essentially segregated from the regions in which the density of the electron fluid is concentrated. Should the ions form a liquid and be free to roam around in the same areas where the electron plasma is located, the electrons would find no difficulty to superimpose to their own oscillations the much slower oscillations necessary to compensate those of the ions, thus effectively screening the e.m. radiative field. All this simply means that, as in the case of ^{4}He, the Born-Oppenheimer approximation can break down only if **the equilibrium positions of the ions'**

oscillations remain fixed, in well defined space-points, where the much lighter electrons cannot completely follow the ions' trajectories. Another, equivalent way to see the necessity of strict spatial localization for the ions' equilibrium positions is to note that in this way only the small fraction of the electrons' modes that have their momenta close to the boundary $\frac{\pi}{a}$ (close to k_F) of the first Brillouin zone, may screen the e.m. interaction. In fact only those modes that have a good spatial overlap with the lattice periodic structure may develop the electrical counteroscillations that cut the e.m. field off. We have thus reached the important conclusion that the "raison d' être", the basic physical foundation for the existence of solids with well defined lattice structure, is just the appearance of coherent configurations of the ions' plasma wave-field in interaction with well defined modes of the e.m. field.

In order to completely characterize such coherent processes, the discussion in Section (5.1) requires two basic inputs ω_R and α_{max}, the maximum amplitude of the harmonic oscillation. To get ω_R we need simply repeat the arguments leading to (5.53) to obtain (again with $a \simeq 2.5\text{Å}$, $m = 50$ Gev, $Q = +1$)

$$\omega_R = 1.38\omega_{Np} = .044\text{eV}, \tag{5.73}$$

implying a CD size

$$\lambda = \frac{2\pi}{\omega_R} \simeq 2.8 \cdot 10^{-3}\text{cm}, \tag{5.74}$$

about 30μ, a typical size of crystal grains.

As for α_{max} we may determine it by estimating the maximum amplitudes of ions' oscillations $\xi_{max} \simeq .25\text{Å}$, which follows from attributing the ion a radius of the order of 1Å.

Using (5.33), we obtain

$$\alpha_{Nmax} = \xi_{max}\left(m\omega_R\right)^{\frac{1}{2}} \simeq 5.87, \tag{5.75}$$

showing that the large oscillation amplitude plasma model of Section (5.1) can be realistically employed. This model has been completely solved (see Eqs. (5.41), (5.42)); the coupling constant $g = 1.05$ implied by (5.73) is well above the coherence threshold .77, and from Fig. 5.1 we determine the gap for the ions' plasma as

$$\delta_{Np} = \omega_R\alpha_{N\ max}^2\frac{[1 + x^2 - 3gx]}{x^2} \simeq 1\,\text{eV}, \tag{5.76}$$

a number that looks very reasonable.

Let us now turn our attention to the ions' wave-field and to the quantum structure of the CGS. From (5.10) the wave-field can be written:

$$\psi_N(\vec{x}, \vec{\xi}; t) = \sum_{\vec{m}\vec{n}} a_{\vec{m}\vec{n}} \psi_{\vec{m}}(\vec{x})_{\vec{m}} \left\langle \vec{\xi} \middle| \vec{n} \right\rangle \tag{5.77}$$

where $\vec{m}$ is a set of integer vectors specifying the lattice sites $\vec{x}_{\vec{m}}$, and

$$\psi_{\vec{m}}(\vec{x}) = \left(\frac{1}{\sqrt{\pi}\sigma}\right)^{\frac{3}{2}} \exp{-\frac{(\vec{x} - \vec{x}_m)^2}{2\sigma^2}}, \tag{5.78}$$

are the normalized wave-functions giving the QM distributions of the equilibrium positions (note that $\sigma \simeq (\frac{1}{m\omega_R})^{\frac{1}{2}} \simeq 4.3 \cdot 10^{-10}$cm, the amplitude of the zero point oscillations of the ions). The quantum operators $a_{\vec{m}\vec{n}}$ obey the commutation relations

$$[a_{\vec{m}\vec{n}}, a^+_{\vec{m}'\vec{n}'}] = \delta_{\vec{m}\vec{m}'}\delta_{\vec{n}\vec{n}'}. \tag{5.79}$$

The solutions of the CE's that describe the CGS, that we have just discussed, imply that in the matter sector the CGS $|\Omega\rangle$ has the following explicit structure:

$$|\Omega\rangle = \prod_{\vec{m}\vec{n}} |\beta_{\vec{m}}\rangle_{\vec{m}} |\alpha_{\vec{n}\vec{m}}\rangle_{\vec{n}\vec{m}}, \tag{5.80}$$

where $|\beta\rangle_{\vec{m}}$ is the coherent state of complex amplitude β (with $|\beta|^2 = 1$) (See Section (1.2)) of the oscillator belonging to the lattice site $\vec{x}_{\vec{m}}$, and $|\alpha\rangle_{\vec{n}\vec{m}}$ is the coherent state with classical complex amplitude α of the quantum oscillator associated to the excited level $\vec{n}$ of plasma oscillations at the site $\vec{x}_{\vec{m}}$. From the solution of the CE's we know that

$$\langle\Omega| \psi_N(\vec{x}, \vec{\xi}; t) |\Omega\rangle = \sum_{\vec{m}} \psi_{\vec{m}}(\vec{x}) \left\langle \vec{\xi} \middle| \alpha\vec{u} \right\rangle, \tag{5.81}$$

where $\left\langle \vec{\xi} \middle| \alpha\vec{u} \right\rangle$ is the $\vec{\xi}$-space wave-function of the coherent oscillator state with amplitude $\alpha\vec{u}$:

$$\left\langle \vec{\xi} \middle| \alpha\vec{u} \right\rangle = \left(\frac{m\omega}{\pi}\right)^{\frac{3}{4}} e^{\frac{(\alpha-\alpha^*)^2}{16}} \exp\left\{-\frac{m\omega}{2}\left(\vec{\xi} - \frac{\alpha\vec{u}}{(2m\omega)^{\frac{1}{2}}}\right)^2\right\}. \tag{5.82}$$

Note that $\left\langle \vec{\xi} \middle| \alpha\vec{u} \right\rangle$ **does not** depend on the lattice site, and is the same **throughout** the plasma. It is straightforward to check that (5.80) satisfies (5.81) provided

$$\alpha_{\vec{n}\vec{m}} = \alpha_{\vec{n}} = \prod_{i=1}^{3} \frac{(\alpha u_i)^{n_i}}{\sqrt{n_i!}} e^{-|\alpha|^2/2}, \tag{5.83}$$

thus the CGS $|\Omega\rangle$ may be finally put in the form

$$|\Omega\rangle = \prod_{\vec{m}\vec{n}} |\beta\rangle_{\vec{m}} |\alpha_{\vec{n}}\rangle_{\vec{m},\vec{n}}, \tag{5.84}$$

where both $\alpha_{\vec{n}}$ and β are $\vec{m}$-(space) independent.

We now analyse a remarkable property of crystal structure and in particular of its ground state $|\Omega\rangle$ [eq. (5.84)]. Let us consider the matrix element

$$F_{\vec{q}} = \langle\Omega| \int_{\vec{x}\vec{\xi}} e^{i\vec{q}\cdot(\vec{x}+\vec{\xi})} \psi_N^+(\vec{x},\vec{\xi};t)\psi_N(\vec{x},\vec{\xi};t) |\Omega\rangle \tag{5.85}$$

i.e. the matter "form factor" of the CGS. Using (5.77) we have

$$\begin{aligned} F_{\vec{q}} &= \langle\Omega| \sum_{\vec{m}\vec{n}\vec{n}'} \int_{\vec{x}\vec{\xi}} e^{i\vec{q}\cdot(\vec{x}+\vec{\xi})} \sum_{\vec{m}} |\psi_{\vec{m}}(\vec{x})|^2 a_{\vec{m}\vec{n}}^+ a_{\vec{m}\vec{n}'} |\Omega\rangle \langle\vec{n}\,|\,\vec{\xi}\rangle \langle\vec{\xi}\,|\,\vec{n}'\rangle \\ &= \sum_{\vec{m}} \int_{\vec{x}\vec{\xi}} |\psi_{\vec{m}}(\vec{x})|^2 e^{i\vec{q}\cdot(\vec{x}+\vec{\xi})} \langle\vec{\alpha}\,|\,\vec{\xi}\rangle_{\vec{m}} \langle\vec{\xi}\,|\,\vec{\alpha}\rangle_{\vec{m}}. \end{aligned} \tag{5.86}$$

In order to perform the integral correctly, let us note that the sum (5.86) adds the form factor of each oscillator whose equilibrium position is the lattice site $\vec{x}_m$. Thus for the oscillator at the generic site $\vec{x}_m$ one may write

$$\langle\vec{\xi}\,|\,\vec{\alpha}\rangle_{\vec{m}} = \int_{\vec{p}} 0_{\vec{m}} \langle\vec{\xi}\,|\,\vec{p}\rangle \langle\vec{p}|\,\vec{\alpha}\rangle_{\vec{m}} = \int_{\vec{p}} \frac{e^{i\vec{p}\cdot(\vec{x}_m+\vec{\xi})}}{(2\pi)^{\frac{3}{2}}} \langle\vec{p}|\,\vec{\alpha}\rangle_{\vec{m}}, \tag{5.87}$$

and substituting in (5.86) we obtain:

$$\begin{aligned} F_{\vec{q}} &= \int d^3p \langle\vec{\alpha}\,|\,\vec{p}+\vec{q}\rangle \langle\vec{p}|\,\vec{\alpha}\rangle \int_{\vec{x}} \sum_{\vec{m}} |\psi_{\vec{m}}(\vec{x})|^2 \\ &= N \exp\Big\{ -\Big(\frac{\vec{q}^2}{2m\omega} - \frac{i}{4}\vec{q}\frac{\vec{\alpha}+\vec{\alpha}^*}{(2m\omega)^{\frac{1}{2}}}\Big)\Big\}. \end{aligned} \tag{5.88}$$

This simple, dramatic consequence of the coherent, collective dynamics of the ions' plasma springs from the simple fact that **all** nuclei have the **same** behaviour, hence the **same** form factor, the phases being the same for **all** ions. Eq. (5.87) is to be contrasted with the usual "incoherent" form factor

$$F_{\vec{q}}^{inc} = \sum_{\vec{m}} e^{i\vec{q}\cdot\vec{x}_m} \tag{5.89}$$

which becomes coherent – $O(N)$ – only when $\vec{q}$ belongs to the reciprocal lattice. $F_{\vec{q}}^{inc}$ lies of course at the basis of all modern quantum crystallography. Eq. (5.88) plays a fundamental rôle in the coherent scattering of different particles' probes off a crystal. However, $F_{\vec{q}}$ as given by (5.88) is not the whole story, for there are other aspects related to quantum and thermal fluctuations that interfere with full coherence. We shall give them some attention in the next Section where we discuss the fascinating phenomenon of the Mössbauer effect.

5.4 The Mössbauer effect and the eclipse of "Asymptotic Freedom"

The Mössbauer effect (ME) was discovered in 1957 by the young German Ph.D. student Rudolf Mössbauer[1], whose work was honoured with the Nobel Prize a few years later. The ME came as a big surprise for, as we shall now see, its physics is so extraordinary that defies common sense. However, after a frenzy that lasted only for a few years, physicists have now committed its subtle phenomenology to chemists, who use it for sophisticated diagnostical studies: the ME rarely appears in physics textbooks on CM, and is usually regarded as a curious example of how strange Quantum Mechanics can be: so damned counterintuitive, isn't it?

In this Section I am going to discuss an analysis of the ME published in collaboration with T. Bressani and E. Del Giudice[2] that will demonstrate that the ME is not an ordinary quantum mechanical effect, but implies the subtle working of electrodynamical coherence in crystals. What is then the ME? For convenience, I shall briefly explain it in the case of a highly popular system -Fe^{57}- instead of the Ir^{191} originally employed by Mössbauer. The relevant nuclear level scheme is depicted in Fig. 5.2, showing that the Mössbauer nuclide is prepared via e-capture from Co^{57}, that populates an excited level – the $\frac{3}{2}^{-}$ at $\omega = 14.37$ keV – much more abundantly than the ground state $\frac{1}{2}^{-}$.

The lifetime of this transition, a γ-transition, is $\tau = 1.4 \cdot 10^{-9}$ sec, corresponding to a width

$$\Gamma \simeq 5 \cdot 10^{-9} \, \text{eV}, \tag{5.90}$$

extremely small for an e.m. decay.

The idea that led to the discovery of the ME was to use this source of narrow, almost monochromatic γ-rays for experiments of nuclear resonance absorption, or, put differently, to shine these γ's from a crystal of Fe^{57} on another Fe^{57} crystal to measure the cross-section for the reverse transition, $\frac{1}{2}^{-} \to \frac{3}{2}^{-}$, induced by the absorption of the γ-ray. In order to check our expectations let us do some kinematics in the case of a very cold environment, so as to avoid the interference of thermal fluctuations. In this case, energy momentum conservation implies that the energy E_{γ} and momentum

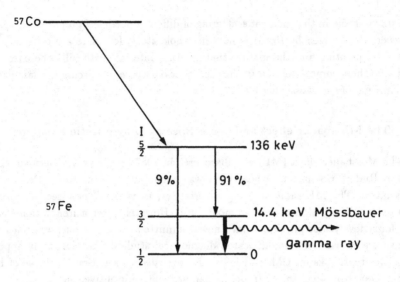

Fig. 5.2. The nuclear level scheme used in the Mössbauer effect of Fe^{57}.

$\vec{k}_\gamma$ of the emitted γ obey the following equations:

$$E_\gamma = \omega - \frac{\vec{k}^2}{2m_{Fe}}, \tag{5.91a}$$

$$\vec{k} = -\vec{k}_\gamma, \tag{5.91b}$$

implying that the emission energy is not the transition energy but a somewhat lower value :

$$E_\gamma^{emis} = \omega - \frac{\omega^2}{2m_{Fe}}, \tag{5.92}$$

as required by the energy that must be converted to the kinetic energy of the recoiling nucleus.

A similar argument shows that the γ-ray can be absorbed at the somewhat higher energy

$$E_\gamma^{abs} \simeq \omega + \frac{\omega^2}{2m_{Fe}}, \tag{5.93}$$

necessary to set the nucleus in motion as demanded by momentum conservation.

The Doppler shifts (5.92) and (5.93) act in such a way that the resonance between emission and absorption goes out of tune by twice the recoil energies:

$$\Delta E = E_\gamma^{abs.} - E_\gamma^{emis} = \frac{\omega^2}{m_{Fe}} \simeq 3 \cdot 10^{-4} \, \text{eV}, \qquad (5.94)$$

about five orders of magnitude larger than the natural width (5.90)! This evidently means that resonance absorption cannot take place at low temperatures, the only possibility for this to happen being the presence of thermal effects that widen the natural width by imparting the decaying nuclei thermal motions that can overcome the detuning (5.94).

This is in fact the opposite of what Mössbauer observed: on decreasing the temperature the resonance absorption cross-section was found to dramatically increase, as though emission and absorption happened without the recoil of the nuclei involved, or better still as though the recoil process involved not the single nucleus but, according to the remarkable measurements of Pound and Rebka, and Craig, Nagle and Cochran[3], a very large number of nuclei, in excess of

$$N_{nucl} \simeq 2 \cdot 10^9, \qquad (5.95)$$

corresponding to volumes almost one micron across!

It was Mössbauer himself who, after the initial puzzlement, showed how a theory developed by Lamb and Dicke could be brought to bear upon his measurements[4]. This theory is still the generally accepted explanation of the ME: it pictures the lattice as an **infinitely stiff** mathematical lattice whose excitations – the phonons – have accordingly a spectrum with momenta $\vec{k}$ limited to the first Brillouin zone ($|\vec{k}| \leq \frac{\pi}{a}$, in a cubic lattice of lattice constant a). It is well known that the discrete translational symmetry of the lattice entails a violation of momentum conservation, the lattice as a whole being able to absorb momenta that belong to the reciprocal lattice. If things are really thus, it can be shown that the quantum state of the lattice and of the nuclei, oscillating around infinitely stiff equilibrium positions, is **completely specified by its phononic, collective excitations.**

The probability that the lattice as a whole can absorb the recoil momentum, explaining in this way the ME, can be calculated in a straightforward fashion. Let us call $\vec{x}_N$ the coordinate of the recoiling nucleus, the probability amplitude for recoilless decay is clearly given by the "form factor"

$$A_{ME} = \exp(-W) = \langle L \mid e^{i\vec{k} \cdot \vec{x}_N} \mid L \rangle \qquad (5.96)$$

with (θ_D is the Debye temperature, m_N the mass of the nucleus)

$$W = \frac{k^2}{2m_N} \frac{1}{\theta_D} \frac{3}{4} \Big[1 + 4\Big(\frac{T}{\theta_D}\Big)^2 \int_0^{\frac{\theta_D}{T}} du \frac{u}{e^u - 1} \Big], \tag{5.97}$$

the Debye-Waller factor. This theory has scored good phenomenological successes, and this together with the high idealization of the model on which it is founded has been a very good motivation to leave the ME at that.

At this point the question one would like to answer is to what extent the idealized mathematical model of Lamb and Dicke is supported by the GACMP, which has never been questioned neither by Mössbauer nor by anybody else. The answer, it must be said very clearly, is that **it is not**, and this rests on a general property of QFT's with short-range interactions that switch off at short distances that may be called "Asymptotic Freedom" (AF). According to AF, when a QFT system is perturbed by perturbations having a momentum spectrum $\vec{k}$ whose support lies much higher than the support of the effective short-range potentials, the system behaves as if it were **free**. And its being arranged in well defined structures at space-scales much larger than the wave lengths of the perturbation is completely immaterial for the dynamics of the interaction matter-perturbation: at those distances the matter system has regained its "freedom".

Let us analyse from this point of view a possible QFT compatible with GACMP. The typical "pseudopotentials" that are employed in the few-body interactions have a momentum spectrum that does not extend much beyond the first Brillouin zone $\frac{\pi}{a} \simeq 1.3 \cdot 10^8 \text{cm}^{-1}$. On the other hand the transition Mössbauer first worked with has $k_\gamma \simeq 6 \cdot 10^9 \text{cm}^{-1}$, corresponding to a wavelength $\lambda \simeq 10^{-9}\text{cm}$, 30 times smaller than the smallest lattice spatial scale, the lattice constant. We **do** have a problem with the AF theorem. Clearly the idealized model, with its infinite stiffness, also violates AF for the simple reason that according to it at distances of the order of $\lambda_\gamma \ll a$ the dynamics of nuclei is "frozen", for the only fluctuations that are allowed by the system are the "phonons", whose momentum spectrum stops with the first Brillouin zone. On the other hand within the GACMP framework at distances $O(\lambda_\gamma)$ the nuclei do fluctuate independently, behaving effectively as **free** particles: to them the simple kinematical arguments that invalidate the ME are perfectly applicable. For GACMP the ME is just impossible!

But the paradox grows thicker when we analyse the times involved in the ME. How long does the ME last? The Heisenberg principle

$$\Delta E \Delta t \geq 2\pi \tag{5.98}$$

gives the answer, provided we know ΔE. Now ΔE is very simply the energy difference between the γ emitted by a free particle and the γ emitted by a particle bound to the lattice, *i.e.* $\Delta E \simeq \frac{\omega^2}{2m_{Fe}}$, which put in (5.98) gives

$$\Delta t_{Moss} \sim \frac{4\pi}{\omega^2} m_{Fe} \simeq 10^{-12} \sec. \tag{5.99}$$

If we now ask ourselves about the distance d travelled by a non e.m. signal like sound (this is all GACMP allows, $v_S \simeq 10^6 \text{cm/sec}$) in the time (5.99) we discover it to be very small: $d \simeq 10\text{\AA}$. But in a sphere of radius d the number N_{nucl} of nuclei ready to "share" the recoil is

$$N_{nucl} \simeq 60 \tag{5.100}$$

some eight orders of magnitude below the lower bound (5.95)!

Thus in GACMP, where AF holds, the idealized model that accounts for the phenomenology of the ME is plainly inconsistent. On the other hand, in a solid where strong ions' coherence holds, the perfect coherence of the short-distance dynamics of nuclei provides precisely that "freezing" of the independent motions of the nuclei that is at the basis of the idealized model. We conclude then that in a crystal governed by the strong coherence of QED, the idealized model is indeed very real, like the ME!

Exercises of Chapter 5

$\langle 5.1 \rangle$: Show why $g_0^2 = \frac{2\pi}{3}$ cannot satisfy both (5.31c) and (5.32)

$\langle 5.2 \rangle$: Perform the estimate leading to (5.63)

$\langle 5.3 \rangle$: Derive Eqs. (5.70) and (5.72).

$\langle 5.4 \rangle$: Go through the steps from (5.80) to (5.88).

References to Chapter 5

1. R.L. Mössbauer, *Z. Phys.* **151** (1958) 124 .
2. T. Bressani, E. Del Giudice and G. Preparata, *Il Nuovo Cimento* **D14** (1992) 345 .

3. R.V. Pound and G.A. Rebka jr., *Phys. Rev. Lett* **4** (1960) 337

    P.P. Craig, D.E. Nagle and D.R.F. Cochran, *Phys. Rev. Lett* **4** (1960) 561

4. For a thorough account of the generally accepted views on ME see H. Frauen-
    felder, *The Mössbauer Effect* (W.A. Benjamin Inc., New York, N.Y., 1962)

Chapter 6

SUPERCONDUCTIVITY, COLD AND HOT

6.1 The kinematics of BCS superconductivity: which "potentials" are good for superconductivity?

Superconductivity (SC), that serendipitous discovery made by Kamerlingh Onnes in 1911, after more than 80 years still fascinates the "natural philosophers" who try to glimpse underneath the marvels and oddities of its boundless phenomenology not only the signs of macroscopic order and organization but its ultimate *raison d'être*, its explanation in terms of the rationality of the laws of nature embodied, nobody doubts it, in QED. For, and this must be affirmed without ambiguity, within GACMP no explanation has ever been given based on the "first principles" of QED, even though an almost endless theoretical literature exists on the most complex and detailed applications to the SC phenomena of a very successful **kinematics**: the Bardeen-Cooper-Schrieffer (BCS) approach[1]. It should be emphasized in fact that BCS superconductivity, successful as it has been since its proposal in 1957, cannot be looked upon as a dynamical theory of the superconductivity of metals at very low temperatures, for the crucial input, the "pairing interaction", has never been given more than impressionistic descriptions and motivations; and these turn out to have no foundation in the interactions of GACMP, as it shall be argued in the course of this Chapter.

To make the analysis as definite and as simple as possible, our starting point is the simple Fermi gas, with spectrum

$$\epsilon_{\vec{p}} = \frac{\vec{p}^{\,2}}{2m},\tag{6.1}$$

that is taken as a model of the conduction electrons of a metal disposed in a half-filled band, whose Fermi momentum is ($\frac{N}{V} = \frac{1}{a^3} \simeq 6.5 \cdot 10^{22}\,\mathrm{cm}^{-3}$, as usual) (see Eq. (5.49))

$$k_F = \frac{(3\pi^2)^{1/3}}{a} \simeq 2.5\,\mathrm{keV}, \tag{6.2}$$

and whose Fermi energy, according to Eq. (6.1), is given by

$$E_F = \frac{k_F^2}{2m} \simeq 6\ \mathrm{eV} \tag{6.3}$$

For the interaction Hamiltonian we take a typical short-range two-body interaction with an effective "pairing potential" $V_{Pair}(\vec{x})$, which we need not specify at this stage. Thus for the Hamiltonian we write

$$H = \int_{\vec{x}} \Psi^\dagger(\vec{x},t) \frac{-\vec{\nabla}^2}{2m} \Psi(\vec{x},t) + \frac{1}{2} \int_{\vec{x},\vec{y}} \Psi^\dagger(\vec{x},t)\Psi(\vec{x},t) V_{Pair}(\vec{x}-\vec{y})\Psi^\dagger(\vec{y},t)\Psi(\vec{y},t), \tag{6.4}$$

where the fermionic wave-field $\Psi(\vec{x},t)$ can be expanded as in (4.77):

$$\Psi(\vec{x},t) = \frac{1}{V^{1/2}} \sum_{\vec{p},s} a_{\vec{p},s}(t)\chi_s e^{i\vec{p}\cdot\vec{x}}, \tag{6.5}$$

and χ_s ($s = +,-$) are the familiar Pauli spinors describing the two spin states of the electron. Substituting (6.5) in (6.4) one readily obtains:

$$H = \sum_{\vec{p},s} \frac{\vec{p}^{\,2}}{2m} a_{\vec{p},s}^\dagger a_{\vec{p},s} + \frac{1}{2V} \sum_{\vec{p}_1,\vec{p}_2,\vec{q},s,s'} (V_{Pair})_{\vec{q}} a_{\vec{p}_1+\vec{q},s}^\dagger a_{\vec{p}_1,s} a_{\vec{p}_2-\vec{q},s'}^\dagger a_{\vec{p}_2,s'} \tag{6.6}$$

where the spin-independent "pairing" potential $((V_{Pair})_{\vec{q}})$ is the Fourier transform of the potential $V_{Pair}(\vec{x})$ appearing in (6.4), i.e.

$$(V_{Pair})_{\vec{q}} = \int_{\vec{x}} e^{-i\vec{q}\cdot\vec{x}} V_{Pair}(\vec{x}). \tag{6.7}$$

We wish now to find the state $|\Omega\rangle$ that minimizes the energy, i.e. the state for which the expectation value of the Hamiltonian operator

$$E = \langle\Omega|\,H\,|\Omega\rangle \tag{6.8}$$

is minimum. The problem can be simplified enormously by approximating the 4-

operator matrix element as

$$\langle \Omega |\, a^\dagger_{\vec{p}_1+\vec{q},s} a_{\vec{p}_1,s} a^\dagger_{\vec{p}_2-\vec{q},s'} a_{\vec{p}_2,s'} \,| \Omega \rangle$$

$$= \langle \Omega |\, a^\dagger_{\vec{p}_1+\vec{q},s} a_{\vec{p}_1,s} \,| \Omega \rangle \langle \Omega |\, a^\dagger_{\vec{p}_2-\vec{q},s'} a_{\vec{p}_2,s'} \,| \Omega \rangle - (\vec{p}_1, s \to \vec{p}_2, s') \qquad (6.9)$$

$$- \langle \Omega |\, a^\dagger_{\vec{p}_1+\vec{q},s} a^\dagger_{\vec{p}_2-\vec{q},s'} \,| \Omega \rangle \langle \Omega |\, a_{\vec{p}_1,s} a_{\vec{p}_2,s'} \,| \Omega \rangle.$$

We note that the "factorization-approximation", that has just been employed, holds for the perturbative vacuum $|0\rangle$ and we expect $|\Omega\rangle$ to deviate only slightly from $|0\rangle$: thus much to justify the adequacy of our approximation. The zero-momentum, zero-angular momentum property of $|\Omega\rangle$ allows us to set:

$$\langle \Omega |\, a^\dagger_{\vec{p},s} a_{\vec{p}\,',s'} \,| \Omega \rangle = \delta_{\vec{p},\vec{p}\,'} \delta_{s,s'} \langle \Omega |\, a^\dagger_{\vec{p},s} a_{\vec{p},s} \,| \Omega \rangle \qquad (6.10a)$$

$$\langle \Omega |\, a_{\vec{p},s} a_{\vec{p}',s'} \,| \Omega \rangle = \delta_{\vec{p},-\vec{p}'} \delta_{s,-s'} \langle \Omega |\, a_{\vec{p},s} a_{-\vec{p},-s} \,| \Omega \rangle = \delta_{\vec{p},-\vec{p}'} \delta_{s,-s'} A_{\vec{p}}, \qquad (6.10b)$$

note that the symmetry of the problem requires $A_{\vec{p},s}$ to be spin independent. It takes a little algebra to cast (6.8) in the form [Ex. $\langle 6.1 \rangle$]:

$$E = \langle \Omega |\, \sum_{\vec{p},s} \{ \mathcal{E}_{\vec{p}} a^\dagger_{\vec{p},s} a_{\vec{p},s} - \frac{1}{2V} \sum_{\vec{q}} (V_{Pair})_{\vec{q}} \left[A_{\vec{p}+\vec{q}} a^\dagger_{\vec{p},s} a^\dagger_{-\vec{p},-s} + A^*_{\vec{p}+\vec{q}} a_{-\vec{p},-s} a_{\vec{p},s} \right] \} \,| \Omega \rangle,$$

$$(6.11)$$

where $\mathcal{E}_{\vec{p}}$ includes all terms multiplying $a^\dagger_{\vec{p},s} a_{\vec{p},s}$. We approach our problem variationally within a class of states

$$| \Omega \rangle = \Pi_{\vec{p},s} (\cos \theta_{\vec{p}} + \sin \theta_{\vec{p}} b^\dagger_{-\vec{p},-s} b^\dagger_{\vec{p},s}) \,| 0 \rangle \qquad (6.12)$$

where $|0\rangle$ is the Fermi liquid ground state, whose energy levels are all filled up to the Fermi energy. In order to avoid an asymmetry between the states with $|\vec{p}| \gtrless p_F$, we have introduced the operators

$$b_{\vec{p},s} = a_{\vec{p},s}, \qquad |\vec{p}| > p_F \qquad (6.13a)$$

$$b_{\vec{p},s} = a^\dagger_{-\vec{p},-s}, \qquad |\vec{p}| < p_F \qquad (6.13b)$$

such that

$$b_{\vec{p},s} \,| 0 \rangle = 0 \qquad (6.13c)$$

for all $\vec{p}$'s.

The physical meaning of (6.12) is quite transparent: to the "free" ground state there is quantum mechanically superposed, with amplitude $\sin \theta_{\vec{p}}$, a pair of "holes" (for $|\vec{p}| < p_F$) or of electrons (for $|\vec{p}| > p_F$) of opposite momenta and opposite spins. $|\Omega\rangle$ describes what in the jargon is referred to as a "condensate of Cooper pairs"[a]. Substituting the definitions (6.13) in (6.11) we obtain (we assume, without prejudice, $A_{\vec{p}}$ real)

$$\Delta E = \langle \Omega | \sum_{\vec{p},s} [\mathcal{E}_{\vec{p}} \delta_{\vec{p}} b^{\dagger}_{\vec{p},s} b_{\vec{p},s} - \frac{1}{2} \Delta_{\vec{p}} (b^{\dagger}_{-\vec{p},-s} b^{\dagger}_{\vec{p},s} + b_{\vec{p},s} b_{-\vec{p},-s})] | \Omega \rangle \qquad (6.14)$$

where $\delta_{\vec{p}} = \pm 1$ for $|\vec{p}| \gtrless p_F$, $\Delta E = E - E_0$ is the difference between E and the energy of $|0\rangle$, and we have introduced the key quantity:

$$\Delta_{\vec{p}} = -\frac{1}{V} \sum_{\vec{q}} (V_{Pair})_{\vec{p}-\vec{q}} A_{\vec{q}}, \qquad (6.15)$$

the gap. We are now going to minimize ΔE subject to the constraint

$$\langle \Omega | \sum_{\vec{p},s} a^{\dagger}_{\vec{p},s} a_{\vec{p},s} | \Omega \rangle = \langle \Omega | \sum_{\vec{p},s} \delta_{\vec{p}} b^{\dagger}_{\vec{p},s} b_{\vec{p},s} | \Omega \rangle + N = N, \qquad (6.16)$$

and this can be done by minimizing the quantity

$$F = \langle \Omega | \sum_{\vec{p},s} [(\mathcal{E}_{\vec{p}} - \mu) \delta_{\vec{p}} b^{\dagger}_{\vec{p},s} b_{\vec{p},s} - \frac{1}{2} \Delta_{\vec{p}} (b^{\dagger}_{-\vec{p},-s} b^{\dagger}_{\vec{p},s} + b_{\vec{p},s} b_{-\vec{p},-s})] | \Omega \rangle, \qquad (6.17)$$

μ is the usual Lagrange multiplier. Putting the explicit form (6.12) in (6.17) we readily obtain ($\epsilon_{\vec{p}} = \mathcal{E}_{\vec{p}} - \mu$):

$$F = \sum_{\vec{p},s} \left[\epsilon_{\vec{p}} \delta_{\vec{p}} \sin^2 \theta_{\vec{p}} - \Delta_{\vec{p}} \sin \theta_{\vec{p}} \cos \theta_{\vec{p}} \right], \qquad (6.18)$$

and equating the derivatives $\frac{\partial F}{\partial \theta_{\vec{p}}}$ to zero yields the following simple relations:

$$\delta_{\vec{p}} \tan 2\theta_{\vec{p}} = \frac{\Delta_{\vec{p}}}{\epsilon_{\vec{p}}}, \qquad (6.19)$$

which allow us to understand at once the meaning of the Lagrange multiplier μ. Continuity for the angles $\theta_{\vec{p}}$, solutions of (6.19), obviously requires that $\mu = E_F$. Around the Fermi surface, where all the action is, $\epsilon_{\vec{p}}$ can be very faithfully approximated by

$$\epsilon_{\vec{p}} = \frac{\vec{p}^{\,2}}{2m} - \frac{\vec{p}_F^2}{2m}. \qquad (6.20)$$

---

[a]Note that, unlike those introduced originally by L. Cooper[2], these "pairs" are completely delocalized in space.

Our problem is solved completely once we determine $\Delta_{\vec{p}}$. This is easily obtained if we recall that

$$A_{\vec{p}} = A_{\vec{p}}^* = \langle \Omega \mid b_{\vec{p},s}b_{-\vec{p},-s} \mid \Omega \rangle = \frac{1}{2}\sin 2\theta_{\vec{p}} = \frac{\Delta_{\vec{p}}}{2\sqrt{\epsilon_{\vec{p}}^2 + \Delta_{\vec{p}}^2}} = \frac{\Delta_{\vec{p}}}{2E_{\vec{p}}}, \qquad (6.21)$$

which substituted in (6.15) yields the all-important integral equation:

$$\Delta_{\vec{p}} = -\frac{1}{2V}\sum_{\vec{q}}(V_{Pair})_{\vec{p}-\vec{q}}\frac{\Delta_{\vec{q}}}{\sqrt{\epsilon_{\vec{q}}^2 + \Delta_{\vec{q}}^2}}, \qquad (6.22)$$

the "gap-equation". This is the end of our kinematical voyage. Before we embark in a discussion of the possible existence of non-trivial solutions of (6.22) (for the trivial solution $\Delta_{\vec{p}} = 0$ always exists, and it does not buy us anything!), thus sailing in the treacherous waters of dynamics, let us briefly discuss the physics of $\mid \Omega \rangle$. What are its "quasi-particle" excitations? This can be readily answered by defining the new operators

$$B_{\vec{p},s} = \cos\theta_{\vec{p}}b_{\vec{p},s} + \sin\theta_{\vec{p}}b_{-\vec{p},-s}^{\dagger} \qquad (6.23a)$$

$$B_{-\vec{p},-s} = \cos\theta_{\vec{p}}b_{-\vec{p},-s} - \sin\theta_{\vec{p}}b_{\vec{p},s}^{\dagger} \qquad (6.23b)$$

obeying the standard anticommutation relations. It is a simple exercise to show that

$$B_{\vec{p},s} \mid \Omega \rangle = 0, \qquad (6.24)$$

i.e. the operators $B_{\vec{p},s}$ annihilate the state $\mid \Omega \rangle$. Furthermore, noting that the operator appearing in the rhs of (6.14) is the leading piece (quadratic) H_q of the Hamiltonian minus E_0 (the energy of the state $\mid 0 \rangle$), we can with little algebra show that ($\hat{N}$ is the number operator)

$$H \simeq H_q = E_F(\hat{N} - N) + \sum_{\vec{p},s}E_{\vec{p}}(B_{\vec{p},s}^{\dagger}B_{\vec{p},s} - \sin^2\theta_{\vec{p}}), \qquad (6.25a)$$

where according to (6.20) and (6.21)

$$E_{\vec{p}} = \sqrt{\epsilon_{\vec{p}}^2 + \Delta_{\vec{p}}^2}. \qquad (6.25b)$$

Finally the difference ΔE between the energy of $\mid \Omega \rangle$ and that of the ground state $\mid 0 \rangle$ is easily calculated from (6.25) to be:

$$\Delta E = -\frac{1}{2}\sum_{\vec{p},s}(E_{\vec{p}} - \epsilon_{\vec{p}}), \qquad (6.26)$$

showing that, within our approximation, $|\Omega\rangle$ is the real ground state.

Eqs.(6.25) fully reveal the nature of $|\Omega\rangle$: within our approximations this state corresponds to a new simple Fermi liquid whose quasi-particle excitations (particles for $|\vec{p}| > p_F$, and "holes" for $|\vec{p}| < p_F$) have a spectrum (6.25b) that for $\epsilon_{\vec{p}} \to 0$, $i.e.$ at the Fermi surface, acquires a non-zero gap $\Delta_{\vec{p}}$. This is superconductivity! In fact in order to excite the conduction electron, that moves around the Fermi surface, the perturbing agent must spend an energy bigger than $\Delta_{\vec{p}}$, which may well turn out to be too large. If this is the case the electrons will move freely without interacting with, and dissipating energy to the lattice: they would behave precisely as Kamerlingh Onnes first saw them behave such a long time ago.

The above chain of deductions is essentially all there is to know about the BCS kinematics of SC: all the fancy developments based on the heavy formalism of QFT, electron self-energy functions, vertex parts, Dyson's equations etc. are, in my opinion, only a screen to hide a great embarrassment that assails the theoretician when, starting from this elegant kinematics, he tries to develop some honest and realistic dynamics.

So, the time has come to analyse the dynamics of superconductivity, which the BCS treatment has elegantly enucleated in the gap equation (6.21), that we find it convenient to first generalize to non-zero temperatures. In order to achieve this we note that the $\vec{q}$-sum in the rhs of (6.21) involves the quantity $A_{\vec{q}}$, defined in (6.10b), that must be evaluated for the state $|\Omega, T\rangle$ at the temperature T.

Noting that the inversion of (6.23) yields:

$$b_{\vec{p},s} = \cos\theta_{\vec{p}} B_{\vec{p},s} - \sin\theta_{\vec{p}} B^{\dagger}_{-\vec{p},-s}, \qquad (6.27\text{a})$$

$$b_{-\vec{p},-s} = \cos\theta_{\vec{p}} B_{-\vec{p},-s} + \sin\theta_{\vec{p}} B^{\dagger}_{\vec{p},s} \qquad (6.27\text{b})$$

for $T \neq 0$ we compute

$$
\begin{aligned}
A_{\vec{p}}(T) &= \langle \Omega, T \mid b_{\vec{p},s} b_{-\vec{p},-s} \mid \Omega, T \rangle \\
&= \cos\theta_{\vec{p}} \sin\theta_{\vec{p}} \langle \Omega, T \mid (1 - B^{\dagger}_{\vec{p},s} B_{\vec{p},s} - B^{\dagger}_{-\vec{p},-s} B_{-\vec{p},-s}) \mid \Omega, T \rangle.
\end{aligned}
\qquad (6.28)
$$

Recalling that from Fermi-statistic

$$\langle \Omega, T \mid B^{\dagger}_{\vec{p},s} B_{\vec{p},s} \mid \Omega, T \rangle = \frac{e^{-E_{\vec{p}}/T}}{1 + e^{-E_{\vec{p}}/T}}, \qquad (6.29)$$

the sought generalization of the gap equation to any temperature is readily derived

[Ex.$\langle 6.2 \rangle$]:

$$\Delta_{\vec{p}}(T) = -\frac{1}{2V} \sum_{\vec{q}} (V_{Pair})_{\vec{p}-\vec{q}} \frac{\Delta_{\vec{q}}(T)}{\sqrt{\epsilon_{\vec{q}}^2 + \Delta_{\vec{q}}^2(T)}} \tanh \frac{E_{\vec{q}}}{2T}. \qquad (6.30)$$

Let us assume, as is experimentally suggested, that the state $|\Omega\rangle$ differs only slightly from $|0\rangle$, and this only around the Fermi sphere $\epsilon_{\vec{p}} = 0$. It is thus convenient to represent all quantities around $\epsilon_{\vec{p}} = 0$ in a simplified form. This can be done by setting $\vec{p} = (p_F + \delta_p)\vec{u}$ ($\vec{u}^2 = 1$) and noting that

$$v_F \delta_p \simeq \epsilon_{\vec{p}}, \qquad (6.31)$$

where $v_F = \frac{p_F}{m}$ is the Fermi velocity. Going from the variables $\vec{p}$ to the variables $(\epsilon, \hat{p} = \frac{\vec{p}}{|\vec{p}|})$, in the isotropic case, that we are now discussing, all physical quantities become functions of ϵ only, the distance in energy from the Fermi sphere (ϵ can clearly have both signs). Finally, going to the continuum limit, for the $\vec{q}$-sums $\sum_{\vec{q}}$ we may write

$$\sum_{\vec{q}} \rightarrow \frac{V}{(2\pi)^3} \int q^2 \, dq \, d\Omega_{\vec{u}} \simeq \frac{V}{(2\pi)^3} \frac{p_F^2}{v_F} \int d\epsilon \int d\Omega_{\vec{u}}, \qquad (6.32)$$

where we have set $q = p_F$, for all functions that are to be integrated will be strongly peaked around the Fermi momentum p_F. With these approximations we rewrite the gap equation

$$\Delta(\epsilon, T) = -\frac{1}{4\pi^2} \frac{p_F^2}{v_F} \int d\epsilon' V_{Pair}(\epsilon - \epsilon') \frac{\Delta(\epsilon', T)}{\sqrt{\epsilon'^2 + \Delta(\epsilon', T)^2}} \tanh \frac{\sqrt{\epsilon'^2 + \Delta(\epsilon', T)^2}}{2T} \qquad (6.33)$$

where

$$V_{Pair}(\epsilon - \epsilon') = \int \frac{d\Omega_{\vec{u}}}{4\pi} V_{Pair}(|\vec{p} - \vec{q}|), \qquad (6.34)$$

and $|\vec{p} - \vec{q}|^2 \simeq 2p_F^2(1 - \cos\theta_{\vec{u}}) + \frac{(\epsilon - \epsilon')^2}{v_F^2}$. As for the relation between the gap $\Delta(\epsilon, T)$ and the pair amplitude $A(\epsilon, T)$, this can be written:

$$\Delta(\epsilon, T) = -\frac{1}{2\pi^2} \frac{p_F^2}{v_F} \int d\epsilon' V_{Pair}(\epsilon - \epsilon') A(\epsilon', T). \qquad (6.35)$$

Eqs. (6.33) and (6.34) completely encapsulate the dynamics of SC once, evidently, an explicit expression for the "pairing" potential is fed in, to be derived from some basic interaction mechanism, which must be chosen from the panoply of interactions that QED offers to Condensed Matter. But before "shopping around" for such an interaction in the GACMP "market", let us gain some insight on the nature and characteristics of $V_{Pair}(\epsilon)$ from what is known experimentally.

For $T < T_c$ - the critical temperature - measurements of the correlation length for pairs of electrons indicate that the support in ϵ of $A(\epsilon) = A(\epsilon, 0)$ is very narrow, on the order of the gap $\Delta(0, 0)$ (a few K). Bearing this in mind a look at (6.35) immediately shows that the support in ϵ of $\Delta(\epsilon, T)$ practically coincides with that of the potential $V_{Pair}(\epsilon)$: the gap $\Delta(\epsilon, T)$ and the pair-amplitude $A(\epsilon, T)$ will have comparable supports only if the potential has also such support. This case is very interesting, for the extension 2Δ of the support in ϵ of the potential is related to its range R in space by the simple relation:

$$R \simeq \frac{2\pi}{\Delta q} = \frac{\pi}{\Delta} v_F, \qquad (6.36)$$

as readily follows from the general properties of Fourier transforms (Heisenberg principle). Choosing the typical value $\Delta \simeq 10$ K, $v_F \simeq \pi/am_e \simeq 5 \cdot 10^{-3}$ one gets $R \simeq 10^{-4}$cm $\simeq 10^4 \mathring{A}$: about ten thousand times the typical ranges (a, the lattice constant) of the electrostatic forces of GACMP! This sounds clearly absurd, and it thus seems inevitable to discard such possibility and to accept as an obvious fact that $\Delta(\epsilon)$ and $A(\epsilon)$ have very different support properties, the support of $\Delta(\epsilon)$ being comparable with that of $V_{Pair}(\epsilon)$.

On the other hand the highly successful phenomenology of BCS superconductivity, as well known, is based upon the momentous BCS Ansatz, that in our notation can be written:

$$V(\epsilon) = \begin{cases} -V_{BCS} & \text{for } |\epsilon| < \omega_D, \\ 0 & \text{for } |\epsilon| > \omega_D, \end{cases} \qquad (6.37)$$

where ω_D is the Debye frequency, the maximum phonon's frequency. The BCS Ansatz, which has been later refined and greatly complicated, without however essentially changing its basic physics, has two main fundamental features: it is attractive in the energy interval $(0, \omega_D)$, and it is negligible outside this interval.

Using (6.37) the gap equation for $\epsilon = 0$ becomes very simply:

$$\Delta(0, T) = +\frac{\lambda}{2} \int_{-\omega_D}^{\omega_D} d\epsilon \frac{\Delta(\epsilon, T)}{\sqrt{\epsilon^2 + \Delta(\epsilon, T)^2}} \tanh \frac{\sqrt{\epsilon^2 + \Delta(\epsilon, T)^2}}{2T} \qquad (6.38)$$

where $\lambda = \frac{1}{2\pi^2} \frac{V_{BCS}}{v_F} p_F^2$ is a positive dimensionless quantity, whose size obviously depends on the positive quantity V_{BCS}. Assuming, as experimentally observed,

$\Delta(0, T) \ll \omega_D$ the integral is dominated by the region $\epsilon \simeq 0$, so we can approximate the integrand by $\Delta(0, T)$, obtaining the standard BCS equation:

$$1 = \frac{\lambda}{2} \int_{-\omega_D}^{\omega_D} d\epsilon \frac{1}{\sqrt{\epsilon^2 + \Delta(\epsilon, T)^2}} \tanh \frac{\sqrt{\epsilon^2 + \Delta(\epsilon, T)^2}}{2T}, \tag{6.39}$$

whose (approximate) solution for $T = 0$ is the well known expression:

$$\Delta(0, 0) \simeq \omega_D e^{-\frac{1}{\lambda}}, \tag{6.40}$$

and for the critical temperature T_c (for which (6.38) yields the solution $\Delta(0, T_c) = 0$) one obtains:

$$2T_c = 1.14\Delta(0, 0). \tag{6.41}$$

Equations (6.40) and (6.41) are the pivotal results of BCS superconductivity, and upon them there has been based an extremely wide and successful phenomenology. Eq. (6.40) exhibits the typical non-perturbative (in λ) nature of the gap at zero temperature, whose proportionality to the Debye frequency correctly accounts for the fundamental "isotope effect": the dependence of $\Delta(0, 0)$ inversely proportional (like ω_D) to the square root of the masses of two, otherwise equivalent, isotopes. (6.41) provides a relation between $\Delta(0, 0)$ and T_c that is quite well satisfied by all low temperature (cold) superconductors. Finally (6.35), implying in general that $\Delta(0, 0) \ll \omega_D$, leads us to expect superconductivity to be a very general phenomenon due to its holding for any attractive pairing interaction ($\lambda > 0$), though sometimes with T_c much too small to observe.

We may thus conclude this brief review of BCS superconductivity with the observation that any theory of superconductivity that wishes to accurately describe the experimental facts of cold superconductivity, must produce a pairing potential that closely resembles the BCS Ansatz. This latter statement underscores the great importance of what has been achieved by BCS: they have for the first time deciphered the deep meaning of the fascinating structure of the superconductive state. But for a "natural philosopher" it is perfectly clear that BCS superconductivity cannot be the whole story, for him the fundamental question really is: from what first principles, i.e. from what QED interactions does a potential of the type (6.37) originate ? I must state with all clarity at this point that the "explanation" that BCS tried to give of (6.37), and for that matter everybody else after them, within GACMP does not have any justification, as we shall now see.

First, let us "naively" Fourier analyse the BCS potential (6.37). The simple arguments that led us to equation (6.35) allow us to estimate the range of the BCS potential (we take a typical $\omega_D \simeq 300$ K)

$$R_{BCS} \simeq \frac{\pi}{\omega_D} v_F \simeq 10^{-5} \text{cm} \simeq 400\text{Å}, \qquad (6.42)$$

which from the standpoint of GACMP is still quite absurd, exceeding the typical ranges of its potentials by almost three orders of magnitude ! So it appears that the analysis cannot be so "naive"; but what can it be ? BCS, and everybody else after them, start from an interesting suggestion by H. Fröhlich[3] that the pairing interaction could well originate from the retarded electron-phonon interaction, whose structure can easily be found to be [Ex.⟨6.3⟩]

$$V_{\vec{q}} \simeq g^2 \frac{\omega_{\vec{q}}}{\left(\epsilon_{\vec{p}} - \epsilon_{\vec{p}+\vec{q}}\right)^2 - \omega_{\vec{q}}^2}, \qquad (6.43)$$

where $\omega_{\vec{q}} \simeq v_s |\vec{q}|$ is the phonon frequency and v_s is the sound velocity (typically $v_s \simeq 10^{-5}$). Inserting (6.43) in equation (6.34) we obtain without much effort the following potential [Ex. (6.4)]

$$V_{ep}(\epsilon) = \frac{g^2}{\omega_D}\left[\frac{\epsilon}{2\omega_D} \log\left|\frac{1 + \frac{\epsilon}{\omega_D}}{1 - \frac{\epsilon}{\omega_D}}\right| - 1\right] \qquad (6.44)$$

which is plotted in Fig. 6.1, showing that it is attractive for $\epsilon < \omega_D$,

repulsive for $\epsilon > \omega_D$, with a repulsive logarithmic singularity precisely at $\epsilon = \omega_D$: the typical behaviour of a dispersive interaction. Note that by taking more "realistic" electron-phonon interactions that allow the coupling constant g to depend the exchange phonon frequency $\omega_{\vec{q}}$ will not substantially change the structure depicted in Fig. 6.1, which can be thus considered as a rather faithful representation of the key features of the realistic electron-phonon interaction. Note also that we leave completely free the strength of g^2, which however plays a fundamental rôle in determining the size of the zero-temperature gap $\Delta(0,0)$. In the literature now comes a remarkable piece of acrobatics, which should make it possible to derive from (6.44) the BCS potential (6.37); the message is: just forget what happens to $V_{ep}(\epsilon)$ for $|\epsilon| > \omega_D$, and set it equal to zero. It is impossible to deny that this is precisely what one should be able to do in order to go from V_{ep} to the BCS potential, but the real question is: why should one cut-off the high ϵ-tail of V_{ep}? To the puzzled student who searches feverishly the literature to get an answer to this fundamental simple question, as far as I could make out, very little satisfaction is offered: all answers he gets are at best complicated exercises in sophistry. Let us review some of these: One main argument goes back to the pioneering analysis of L. Cooper[2] to argue that the Pauli-principle constrains the pair-wave function $A(\epsilon, 0)$ to have its support concentrated in a narrow

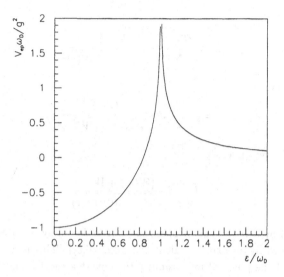

Fig. 6.1. The Fröhlich electron-phonon potential as a function of $x = \frac{\epsilon}{\omega_D}$

interval (of width 2Δ) and this makes the electron-phonon potential inoperative for $|\epsilon| \gg \Delta$, so that cutting it off for $|\epsilon| > \Delta$ ($\omega_D \gg \Delta$) should be completely safe. This argument is clearly completely wrong for the gap equation is not for $A(\epsilon, T)$ but for $\Delta(\epsilon, T)$ and, as we have seen, (6.35) is there to tell us that $\Delta(\epsilon, T)$ has the same support of $V(\epsilon)$. As a matter of fact if $A(\epsilon, T)$ is so narrowly distributed, setting $W(\epsilon) = \frac{V(\epsilon)}{|V(0)|}$, ($W(0) = -1$) Eq. (6.35) allows us to write with good accuracy (we now drop the temperature variable)

$$\Delta(\epsilon) \simeq -\Delta(0)W(\epsilon). \tag{6.45}$$

In this case the mathematical nature of the problem of the gap-equation is completely different from the one resulting from the BCS potential (6.37). In fact, substituting (6.45) in (6.33) for $T = 0$ one gets (g is a positive constant):

$$W(\epsilon) = -g \int_{-E_F}^{E_F} d\epsilon' \frac{W(\epsilon - \epsilon')W(\epsilon')}{\sqrt{\epsilon'^2 + \Delta(0)^2 W(\epsilon')^2}} \tag{6.46}$$

an integral equation for the "normalized" potential $W(\epsilon)$.

It is not difficult now to see that for potentials of the type (6.44) (6.46) has no solution: for instance if we look at (6.46) for $\epsilon = 0$ and for ϵ large, in the former case one has:

$$1 = g \int_{-E_F}^{E_F} d\epsilon \frac{W(\epsilon)^2}{\sqrt{\epsilon^2 + \Delta(0)^2 W(\epsilon)^2}} \tag{6.47a}$$

while in the latter:

$$-1 = g \int_{-E_F}^{E_F} d\epsilon \frac{W(\epsilon)}{\sqrt{\epsilon^2 + \Delta(0)^2 W(\epsilon)^2}} \tag{6.47b}$$

and summing the two gives:

$$0 = \int_{-E_F}^{E_F} d\epsilon \frac{[W(\epsilon)^2 + W(\epsilon)]}{\sqrt{\epsilon^2 + \Delta(0)^2 W(\epsilon)^2}} \tag{6.47c}$$

which is obeyed by a BCS type of potential, but is definitely violated by (6.44), as can be ascertained by an explicit calculation [Ex. ⟨6.5⟩]. This inconsistency is just another signal of the physical impossibility to convert a short-range potential, such as the electron-phonon, into one that in fact "effectively" has the much larger range (6.42).

Another argument attributes the obligatory cut-off at $|\epsilon| = \omega_D$ to some "subtle" interference between Coulomb repulsion and electron-phonon attraction, concluding that the repulsive part of the latter (for $|\epsilon| > \omega_D$) can be absorbed (?!), thus neglected, in the former that is always repulsive. In this way an interaction that should, and in fact does, hamper SC becomes the "deus ex machina" that rescues an otherwise impossible SC!

Usually good, sound physics does not make us expect this kind of miracles to happen. So the firm conclusion of this latter discussion appears to be that within GACMP the "natural philosopher" has indeed found no believable physical answer to his wondering about the mysteries and the marvels of superconductivity: no mechanism has been found that can boost the short ranges of the GACMP potential to the (relatively) huge distances of R_{BCS}.

In Section (6.3) we shall see that a completely different situation arises when quantum electrodynamical coherent mechanisms will be explored.

6.2   Understanding ^{3}He superfluidity

In Section (4.4) we have described the "weak coherence" process involving the electrons of the liquid ^{3}He, that oscillate between the $1S$ and nP configurations of parahelium, in phase with a "classical" electromagnetic field that, as a consequence, remains trapped in the system. At the end of that Section it was argued that the electrons' weak coherence may generate collective (over a CD, whose size is about 600Å) angular momenta of non-negligible amplitude (see equation (4.94)), if by doing so the global system can gain energy. Furthermore a mechanism was identified in Section (4.4) that could precisely induce the emergence of such angular momenta, thus providing a pair interaction among the nuclear spins of ^{3}He that could lead to an explanation of the strong anisotropies of ^{3}He superfluidity.

The starting point of our analysis[4] is thus the hyperfine interaction between the spin vector ($|\vec{s}| = \frac{1}{2}$) of the ^{3}He-nucleus and the electrons' angular momentum $\vec{L}_e$. The Hamiltonian (4.96) for the wave-field of ^{3}He-nuclei can be written:

$$H_{hf}(t) = g_{hf} \int_{\vec{x}} \Psi^\dagger(\vec{x},t) \frac{\vec{\sigma}}{2} \Psi(\vec{x},t) \cdot \vec{L}_e(\vec{x}), \qquad (6.48)$$

where experimentally the hyperfine coupling constant is

$$g_{hf} = 3.18 \cdot 10^{-3} \, \text{cm}^{-1}, \qquad (6.49)$$

$\Psi(\vec{x},t)$ is the usual spin $\frac{1}{2}$-wave-field and the electrons' angular momentum is given by (see Eq. (4.94))

$$(L_e)_i = N_{CD} \theta^2 i \epsilon_{kih} u_k^* u_h |f(\vec{x})|^2, \qquad (6.50)$$

where $f(\vec{x})$ $[f(\vec{0}) = 1]$ is a periodic function with period $\sqrt{2}\frac{2\pi}{\omega_0}$, (see Section (3.5)) $\omega_0 = 21 \, eV$, the fundamental frequency of electrons' e.m. coherence. Note that by choosing $N = N_{CD}$ we are assuming that the nuclei interact with the "resultant" angular momentum of a full Coherence Domain, for this is the minimum unit in which electrons are coherent, and the "weak coherence" of the process under consideration does not make it reasonable to sum the angular momenta in regions which extend much beyond a single CD. Also, the coherent sums of electrons' angular momenta, a well known phenomenon in the case of nuclear spins [recall Nuclear Magnetic Resonance (NMR)] makes $|\vec{L}_e|$ sizeable even though θ is quite small ($N_{CD}\theta^2 \simeq 50$).

We compute the pairing potential [see Fig. 6.2] by second order time dependent perturbation theory in $H_{hf}(t)$ ($|\Omega\rangle$ is the ground state we search for) which we write

$$V_{\vec{p}\,'r's',\vec{p}rs} =$$

$$= -\frac{i}{2} \int_{-\infty}^{\infty} dt' \frac{\langle\Omega| \, a_{-\vec{p}',-s'}\left(\frac{t}{2}\right) a_{\vec{p}',r'}\left(\frac{t}{2}\right) T\left(H_{hf}\left(\frac{t}{2}\right) H_{hf}\left(\frac{-t}{2}\right)\right) a_{\vec{p},r}^\dagger\left(\frac{-t}{2}\right) a_{-\vec{p},-s}^\dagger\left(\frac{-t}{2}\right) |\Omega\rangle}{\langle\Omega| \, a_{-\vec{p},s}^\dagger a_{\vec{p},r} a_{\vec{p},r}^\dagger a_{-\vec{p},s}^\dagger |\Omega\rangle}$$

$$(6.51)$$

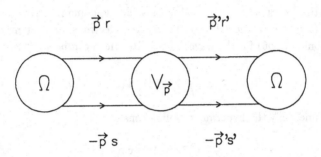

Fig. 6.2. The diagram describing the pairing potential in the non-perturbative ground-state $|\Omega\rangle$.

where the denominator stems from the lack of normalization of the pair wave-function $a^\dagger_{\vec{p},r} a^\dagger_{-\vec{p},s} |\Omega\rangle$. Summing over a complete set of intermediate states, and recalling that the "free" particle operators $a_{\vec{p},r}(t)$ evolve in time according to the "free" Hamiltonian i.e. $a_{\vec{p},r}(t) = e^{-i\epsilon_{\vec{p}} t} a_{\vec{p},r}(0)$ it is a straightforward task to obtain for (6.51)

$$V_{\vec{p}\,'r's',\vec{p}rs} \simeq \frac{g^2_{hf}}{V} \langle L_i L_j \rangle \left(\frac{\sigma_i}{2}\right)_{r'r} \left(\frac{\sigma_j}{2}\right)_{s's} \frac{2}{\epsilon_{\vec{p}} + \epsilon_{\vec{p}\,'} - E_{\vec{p}} - E_{\vec{p}\,'}} \int_{\vec{x}} |f(\vec{x})|^2 e^{i(\vec{p}-\vec{p}\,')\cdot\vec{x}} \quad (6.52)$$

where, as usual, $E_{\vec{p}} = \sqrt{\epsilon^2_{\vec{p}} + \Delta^2_{\vec{p}}}$.

The energy denominator of (6.52) ($\epsilon_{\vec{p}}$ is, we recall, the energy distance of the free states from the Fermi surface) is noteworthy: it stems from our summing over physical states, whose spectrum has acquired a gap. Note that it is always negative, i.e. the pairing potential is always attractive. Approximating $f(\vec{x}) \simeq \frac{\sin \omega_0 r}{\omega_0 r}$ the "form factor" can be computed explicitly:

$$\int_{\vec{x}} |f(\vec{x})|^2 e^{i(\vec{p}-\vec{p}\,')\cdot\vec{x}} = \frac{\pi^2}{\omega_0^2 |\vec{p} - \vec{p}\,'|} \theta(-|\vec{p} - \vec{p}\,'| + 2\omega_0). \quad (6.53)$$

Let us now turn our attention to the spin-dependent part, and in particular to the average $\langle L_i L_j \rangle$. We have

$$\langle L_i L_j \rangle = -\epsilon_{hik} \epsilon_{mjn} \langle u^*_n u_k u^*_m u_n \rangle (N_{CD} \theta^2)^2, \quad (6.54)$$

where the average stems from the fact that the unit vectors u_k may vary from one CD to another being fixed only for intra-CD correlations. We may thus write

$$\langle L_i L_j \rangle = \frac{2}{9}(N_{CD}\theta^2)^2 \delta_{ij} \qquad \text{for inter-CD correlations} \qquad (6.55a)$$

$$\langle L_i L_j \rangle = (N_{CD}\theta^2)\hat{L}_i\hat{L}_j \qquad \text{for intra-CD correlations.} \qquad (6.55b)$$

The importance of the distinction between inter-CD and intra-CD correlation can be argued as follows. In the case of dominance of inter-CD correlations, which should happen at very low temperatures when the Fermi liquid states are spread in space, the pairing potential is a negative function of $\epsilon_{\vec{p}}$ and $\epsilon_{\vec{p}'}$ multiplying a term of the type $\vec{S}_1 \cdot \vec{S}_2$. As the gap function $(\Delta_{\vec{p}})_{rs}$ must be an eigenfunction of the potential and must be odd for the interchange of the two fermions ($\vec{p} \to -\vec{p}$, $r \longleftrightarrow s$) (from the Pauli principle), the pairing potential is attractive in the triplet state $|\vec{S}_1 + \vec{S}_2| = 1$ while in the singlet state $|\vec{S}_1 + \vec{S}_2| = 0$ it is repulsive. It thus turns out that the gap $(\Delta_{\vec{p}})_{rs}$ must have just the form proposed by Balian and Werthamer (BW)[5]:

$$(\Delta_{\vec{p}})_{rs} = \sqrt{\frac{4\pi}{3}}\Delta(|\vec{p}|)\sum_m Y_1^m(\hat{p}) \langle rs \mid 1, m \rangle \qquad (6.56)$$

where $Y_1^m(\Omega)$ are the well known spherical harmonics and $\langle rs \mid 1, m \rangle$ is the spin-1 wave-function for the two spinors. Eq. (6.56) accounts well for the properties of the superfluidity at low temperatures: *i.e.* in the B-phase.

If, on the other hand, the intra-CD correlations dominate, and this may happen when some thermal disorder has begun to develop in the interstices between CD's (so that the longest correlations are within a single CD), Eq. (6.55b) produces the anisotropic interaction $(\hat{L} \cdot \vec{S}_1)(\hat{L} \cdot \vec{S}_2)$ and the anisotropic gap (we choose $\hat{L}$ in the z-direction)

$$(\Delta_{\vec{p}})_{rs} = \sqrt{\frac{4\pi}{3}}\Delta(|\vec{p}|)\left[Y_1^1(\hat{p}) \langle rs \mid 1, 1 \rangle + Y_1^{-1}(\hat{p}) \langle rs \mid 1, -1 \rangle\right] \qquad (6.57)$$

which has precisely the form proposed by Anderson, Brinkman and Morel (ABM)[6], that is known to correctly account for the A-phase. So far, so good!

Putting (6.56) in the spin-dependent gap equation, we obtain after some algebra the scalar gap equation:

$$\Delta(\epsilon, T)_{BW} = \frac{\lambda_{BW}}{2} \int d\epsilon' \theta\left(\omega_0 - \frac{|\epsilon' - \epsilon|}{2v_F}\right) \left[1 - \frac{|\epsilon' - \epsilon|}{2\omega_0 v_F}\right]$$

$$\cdot \frac{1}{E(\epsilon) + E(\epsilon') - (\epsilon + \epsilon')} \frac{\Delta(\epsilon, T)_{BW}}{E(\epsilon')} \tanh \frac{E(\epsilon')}{2T}, \tag{6.58}$$

where

$$\lambda_{BW} = \frac{2}{9} \frac{g_{hf}^2 (\theta^2 N_{CD})^2}{\omega_0 v_F}, \tag{6.59}$$

and $E(\epsilon) = \sqrt{\epsilon^2 + \Delta(\epsilon)^2}$. The most remarkable feature of (6.58) is that a BCS-type potential has clearly emerged, for the θ-function cuts the integral off at

$$2\omega_0 v_F = 100 \text{ mK}, \tag{6.60}$$

a good "Debye temperature" for ^{3}He, which becomes superfluid at about 1mK. Such a large range for the potential doesn't surprise us at all for it corresponds to the size of the CD that is about 600Å, the typical correlation length of superfluid ^{3}He. Finally by factorizing $\Delta(\epsilon, T)$ out of the gap equation we obtain:

$$1 = \frac{\lambda_{BW}}{2} \int_{-\omega_0 v_F}^{\omega_0 v_F} \frac{d\epsilon}{\sqrt{\epsilon^2 + \Delta(\epsilon)^2}} \frac{1}{\Delta_{BW} + \sqrt{\epsilon'^2 + \Delta_{BW}^2} - \epsilon'} \tanh \frac{\sqrt{\epsilon^2 + \Delta_{BW}^2}}{2T}. \tag{6.61}$$

For the A-phase it is the gap function (6.57) that must be employed, noting that

$$\Delta_{\vec{p}}^2 = (\Delta_{\vec{p}})_{rs}^* (\Delta_{\vec{p}})_{rs} = \Delta(|\vec{p}|)^2 \sin^2 \theta = \Delta(\epsilon, T)_{ABM}^2 \tag{6.62}$$

is now anisotropic. Through some simple computations one readily derives the A-phase gap equation:

$$1 = \frac{\lambda_{ABM}}{2} \int_{-\omega_0 v_F}^{\omega_0 v_F} d\epsilon \frac{1}{2} \int_{-1}^{1} d\cos\theta \frac{1}{E(0, \theta) + E(\epsilon', \theta) - \epsilon} \frac{\tanh \frac{E(\epsilon, \theta)}{2T}}{E(\epsilon, \theta)} \tag{6.63}$$

where

$$\lambda_{ABM} = \frac{1}{4v_F \omega_0} g_{hf}^2 (N_{CD} \theta^2)^2, \tag{6.64}$$

and

$$E(\epsilon, \theta) = \sqrt{\epsilon^2 + \Delta_{ABM}^2} \tag{6.65}$$

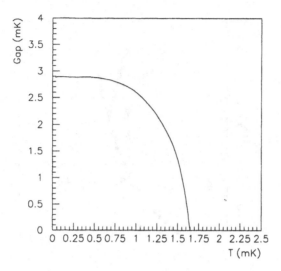

Fig. 6.3. The gap Δ_{BW} as a function of T at zero pressure.

are the anisotropic energies of the quantum fluctuations around the A-phase super-conducting ground-state.

Eqs. (6.61) and (6.63) have been solved numerically[4] yielding the gap Δ_{BW} as a function of T at zero pressure and the $p - T$ phase plane reported in Fig. 6.3 and Fig. 6.4 respectively.

In view of the essentially parameter-free nature of our computation the agreement with experiment is to be judged very satisfactory.

As for the A-phase and the solution of the gap equation for the ABM-wave-function, we have just argued that this latter configuration can be accessed when a substantial amount of thermal fluctuations has been created in the system so as to effectively decouple the CD's. In this case however we expect (6.64) to be renormalized for we cannot pretend that either N_{CD} or θ^2 be the same as in the low temperature phase: a reduction of λ_{ABM} from (6.64) is predicted and found. Fig. 6.4 represents a calculation of the $p - T$ phase plane based on the gap equations (6.61) and (6.63). We could now go on and discuss a number of interesting things about the A-phase especially in presence of magnetic fields, but, it should be known by now, this is not the purpose of our exposition. This Section will end with a brief summary and discussion.

The fact that ^{3}He becomes superfluid at very low temperatures and that its su-perfluidity involves the nuclear spins, a system that is normally coupled very weakly,

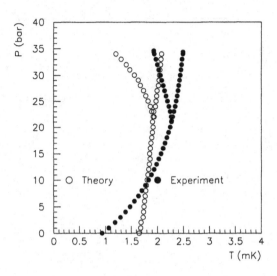

Fig. 6.4. The P-T phase plane of the superfluid ^{3}He from Ref. [4]

is evidently a deep mystery for GACMP. It is thus most rewarding that ^{3}He superfluidity finds a very natural explanation in the hyperfine coupling of nuclear spins with
a coherent electrons' orbital magnetic moment induced by the same "weak coherence" that is responsible for the Superfluidity of ^{3}He. The "spin-orbit" nature of the
hyperfine interaction is therefore the fundamental reason underneath the "P-wave"
gaps (6.56) and (6.57) that characterize the B- and A-phases respectively. Beyond
the numerical successes what is most satisfying about the description afforded by the
electrodynamical coherence is that a phenomenon, that to the "realists" among the
GACMP scientists **must** appear as impossible (for lack of any believable mechanism
possessing the long range and the non-negligible strength of the "observed" pairing
interaction), can, as a matter of fact, be correctly predicted from the **macroscopic**
coherence of the quantum electrodynamical interaction. It really looks as though the
"book of nature" is losing one of its seals.

6.3   Ideas for a theory of SC, cold and hot

In Section (6.1) our analysis of the elegant BCS kinematics has revealed the "un-
conventional" nature of the BCS Ansatz (6.37) for the pairing potential of SC, that
is phenomenologically so successful.

We have also seen that neither the Fröhlich's electron-phonon interaction (6.43) nor any other more or less sophisticated (or rather sophistic?) modification of it are capable to overcome the basic problem of the energy cut-off ω_D, whose "naive" implication is a fantastic and irrealistic amplification (see (6.42)) of the expected and reasonable short-range of the fundamental electron-phonon interaction. In the following Section (6.2) it has been shown that the superfluidity of liquid ^{3}He at very low temperatures, which is physically akin to SC (and poses the same kind of problems to GACMP) can be satisfactorily accounted for by the long-range coherent interaction mechanisms that only belong to the radiation field of QED.

Let us recall that in GACMP such mechanisms are excluded from playing any rôle in the interactions among the elementary systems of condensed matter due to some fallacious arguments that have been exposed in Section (3.3). In this Section I shall outline and explain some general ideas, based on electrodynamical coherent interaction mechanisms, that might be of some relevance for finally understanding cold SC and the more recent and puzzling high T_c superconducting ceramic materials. In the spirit of this book, no attempt will be made to cover the field, the discussion will be concentrated solely on ideas: on visualizing the problem in the new theoretical framework.

As already emphasized a central feature of cold SC is the "isotope effect", whose embodiment in the BCS potential has led to the universal acceptance of the BCS approach to SC (and to its being honoured by a Nobel Prize). Within GACMP the electron-phonon interaction appears to be the only one that, possessing the scale ω_D (the isotopic mass behaviour of the Debye energy is, as well known, $\omega_D \sim m^{-1/2}$), is capable to "explain" the effect, and this has led, *faute de mieux*, to the equally universal acceptance of this interaction as the fundamental basis for cold SC. The physical impossibility of this view has been already argued in the preceding Section (6.2); however, it does not seem a good strategy to outright dismiss such views, without trying to save some of the motivations and expectations on which they were based. And the point that cold SC must have something to do with the coupling between the conduction electrons and the vibrations of the nuclei of the solid state lattice is certainly a very good one, after all the isotopic mass is a property that does not belong to any other actor of the dynamics of condensed matter.

If the interaction with phonons – the quantum fluctuations of the ions' wave-field introduced and discussed in Section (5.3) - will not do, for the reasons explained in the preceding Section, why not try the coherent vibration of the wave-field, that we know from Section (5.3) are the very basis of crystal structure ? In fact, we are already guaranteed that no trouble will arise with the range of the pairing potential for the range of QED coherence extends over the size of a CD, if the order of $\frac{2\pi}{\omega_N} \simeq 2.8 \cdot 10^{-3}$ cm (see equation (5.74)). Furthermore the plasma nature of the ions' oscillations, discussed in Chapter 5, leads us to expect the fundamental frequency of oscillation

ω_N to be given by a "renormalized" plasma frequency that one can write

$$\omega_N = \lambda \frac{Z_{eff}e}{m_N^{1/2}}\left(\frac{N}{V}\right)^{1/2}, \tag{6.66}$$

($\lambda \simeq 1.4$ is a renormalization constant (see Eq. (5.53)) and Z_{eff} is the effective charge of the ion) showing the correct isotopic mass dependence.

Our starting point is the interaction Hamiltonian H_I, between the wave-field $\Psi_e(\vec{x}, t)$ of the conduction electrons and the ions' wave-field $\Psi_N(\vec{x}, \vec{\xi}; t)$, defined in Chapter 5, which we write as

$$H_I(t) = \int_{\vec{x}\vec{y}\vec{\xi}} \Psi_N^\dagger(\vec{x}, \vec{\xi}; t)\Psi_N(\vec{x}, \vec{\xi}; t)V_C(\vec{x} + \vec{\xi} - \vec{y})\Psi_e^\dagger(\vec{y}, t)\Psi_e(\vec{y}, t) \tag{6.67}$$

where, for simplicity, the potential $V_C(\vec{x})$ is the attractive screened Coulomb interaction between ions and electrons

$$V_C(\vec{x}) = -\frac{Z_{eff}e^2}{4\pi}\frac{e^{-\mu_D|\vec{x}|}}{|\vec{x}|} = -Z_{eff}e^2 \int \frac{d^3\vec{q}}{(2\pi)^3}\frac{e^{i\vec{q}\cdot\vec{x}}}{\vec{q}^2 + \mu_D^2} \tag{6.68}$$

$\frac{1}{\mu_D}$ is the Debye screening length, of the order of 10^{-8} cm, that in concrete cases can be evaluated by the methods of GACMP.

Expanding as usual the electron wave-field in its plane-wave modes $(\vec{p}, s)$ and substituting (6.68) in (6.67) we have

$$H_I(t) = \sum_{\vec{p}\vec{p}\,',s} \frac{a_{\vec{p}\,',s}^\dagger a_{\vec{p},s}}{V}\frac{-Z_{eff}e^2}{(\vec{p}-\vec{p}\,')^2 + \mu_D^2}\int_{\vec{x}\vec{\xi}} \Psi_N^\dagger(\vec{x}\vec{\xi}; t)\Psi_N(\vec{x}, \vec{\xi}; t)e^{i(\vec{p}-\vec{p}\,')\cdot(\vec{x}+\vec{\xi})}. \tag{6.69}$$

Let us analyse in detail the integral appearing in (6.69). Neglecting, due to its irrelevance, the crystal structure of the wave-function (we are sandwiching the wave-field with its CGS) we may write

$$\Psi_N(\vec{x}, \vec{\xi}; t) = \left(\frac{N}{V}\right)^{1/2} f(\vec{x})\left\langle \vec{\xi} \mid \vec{\alpha}(t) \right\rangle, \tag{6.70}$$

and substituting in the integral, where we have subtracted the screened Coulomb interaction (that has already been included in the electrons' Fermi liquid dynamics), we have

$$\int_{\vec{x}\vec{\xi}} \Psi_N^\dagger(\vec{x}, \vec{\xi}; t)\Psi_N(\vec{x}, \vec{\xi}; t)e^{i(\vec{p}-\vec{p}\,')\cdot\vec{x}}[e^{i(\vec{p}-\vec{p}\,')\cdot\vec{\xi}} - 1]$$

$$\simeq \left(\frac{N}{V}\right)F(\vec{p}-\vec{p}\,')[i(\vec{p}-\vec{p}\,')_i \left\langle \vec{\alpha} \mid \xi_i(t) \mid \vec{\alpha}\right\rangle], \tag{6.71}$$

where for the "form factor" we have (see equation (6.53))

$$F(\vec{q}) = \int_{\vec{x}} |f(\vec{x})|^2 e^{i\vec{q}\cdot\vec{x}} \simeq \frac{\pi^2}{\omega_N^2 q} \theta(2\omega_N - q). \qquad (6.72)$$

Note that $|\alpha\rangle$ is a coherent state.

Defining

$$G(t) = \left(\langle\vec{\alpha}| \, \xi_i(\tfrac{t}{2})\xi_j(\tfrac{-t}{2}) \, |\vec{\alpha}\rangle + \langle\vec{\alpha}| \, \xi_i(\tfrac{t}{2}) \, |\vec{\alpha}\rangle \langle\vec{\alpha}| \, \xi_j(\tfrac{-t}{2}) \, |\vec{\alpha}\rangle \right) \qquad (6.73)$$

the pairing potential, like in the case of ^{3}He superfluidity, can be written as

$$(V_{Pair})_{\vec{p}s;\vec{p}'s'} =$$

$$= -\frac{i}{2}\delta_{ss'} V \int_{-\infty}^{\infty} dt \int_{\vec{x}} \frac{\langle\Omega| \, a_{-\vec{p}',-s}a_{\vec{p}',s}T(H_I(\tfrac{t}{2},\vec{x})H_I(-\tfrac{t}{2},\vec{0}))a_{\vec{p},s}^{\dagger}a_{-\vec{p},s}^{\dagger} \, |\Omega\rangle}{\langle\Omega| \, a_{-\vec{p},-s}a_{\vec{p},s}a_{\vec{p},s}^{\dagger}a_{-\vec{p},-s}^{\dagger} \, |\Omega\rangle}$$

$$= -\frac{i}{2}\frac{2}{V} \left(\frac{Z_{eff}e^2}{(\vec{p}-\vec{p}')^2 + \mu_D^2} \right)^2 \left(\frac{N}{V} \right)^2 (\vec{p}-\vec{p}')_i(\vec{p}-\vec{p}')_j F(\vec{p}-\vec{p}')$$

$$\cdot \left[\int_0^{\infty} dt \, e^{i(\epsilon_{\vec{p}}+\epsilon_{\vec{p}'}-E_{\vec{p}}-E_{\vec{p}'})t}G(t) + \int_{-\infty}^0 dt \, e^{-i(\epsilon_{\vec{p}}+\epsilon_{\vec{p}'}-E_{\vec{p}}-E_{\vec{p}'})t}G(-t) \right]. \qquad (6.74)$$

The structure of (6.74) can be best understood by looking at the two-diagrams in Fig. 6.5, where the crosses represent the interaction between the electron and the ions' field. The energy of the "free" pair state on the real ground state $|\Omega\rangle$ is, according to perturbation theory, the "free" energy $2\epsilon_{\vec{p}}$, whereas the intermediate state must be a physical pair state, whose energy, we know, is $E_{\vec{p}} + E_{\vec{p}'}$ $(E_{\vec{p}} = \sqrt{\epsilon_{\vec{p}}^2 + \Delta_{\vec{p}}^2})$, and this explains the time-exponentials.

In order to complete the calculation of the pairing potential, which now is obviously spin-independent, we need compute

$$\langle\vec{\alpha}| \, \xi_i(\tfrac{t}{2})\xi_j(-\tfrac{t}{2}) \, |\vec{\alpha}\rangle = \langle\vec{\alpha}| \, \xi_i(t)\xi_j(0) \, |\vec{\alpha}\rangle$$

$$= \frac{1}{2m_N\omega_N} \langle\vec{\alpha}| \, (a_i(t) + a_i^{\dagger}(t))(a_j(0) + a_j^{\dagger}(0)) \, |\vec{\alpha}\rangle. \qquad (6.75)$$

Using the resolution of the identity [Eq. (1.55)] (6.75) becomes

$$\langle\vec{\alpha}| \, \xi_i(\tfrac{t}{2})\xi_j(\tfrac{-t}{2}) \, |\vec{\alpha}\rangle = \frac{1}{2m_N\omega_N} \int \frac{d^2\vec{\beta}}{(i\pi)^3} \langle\vec{\alpha}| \, (a_i(t) + a_i^{\dagger}(t)) \, |\vec{\beta}\rangle \langle\vec{\beta}| \, (\beta_j^* + \alpha_j) \, |\vec{\alpha}\rangle. \qquad (6.76)$$

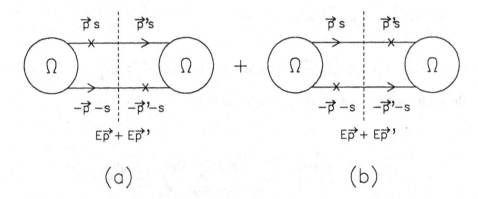

Fig. 6.5. The diagrammatic expression of the electron pairing potential (6.74)

The term proportional to α_j can be immediately integrated to obtain (recall that from Section (5.1) $\langle \vec{\alpha} | a_i(t) | \vec{\alpha} \rangle = \alpha u_i e^{-i\omega_r t}$)

$$\alpha_j \langle \vec{\alpha} | (a_i(t) + a_i^\dagger(t)) | \vec{\alpha} \rangle = \alpha^2 u_j (u_i e^{-i\omega_r t} + u_i^* e^{i\omega_r t}) \qquad (6.77)$$

where $\omega_r = \omega_N (1 - \dot{\phi})$ according to (5.43a) is negative.

In order to compute the term proportional to β_j^* we need analyse

$$\langle \vec{\alpha} | (a_i(t) + a_i^\dagger(t)) | \vec{\beta} \rangle = \langle \vec{\alpha} | e^{iHt}(a_i(0) + a_i^\dagger(0)) e^{-iHt} | \vec{\beta} \rangle =$$
$$= \langle \vec{\alpha} | e^{iHt}(a_i(0) + a_i^\dagger(0)) e^{-iH_0 t} | \vec{\beta} \rangle , \qquad (6.78)$$

for $| \vec{\beta} \rangle$ is a generic intermediate state which, differently from $| \vec{\alpha} \rangle$, in general is out of phase with the electromagnetic field and it evolves according to the "free" Hamiltonian $H_0 = \omega_N a_i^\dagger a_i$. This important observation allows us to write (6.78) as

$$\langle \vec{\alpha} | e^{iHt}(a_i(0) + a_i^\dagger(0)) e^{-iH_0 t} | \vec{\beta} \rangle = \langle \vec{\alpha} e^{-i\omega_r t} | (\alpha_i^* e^{i\omega_r t} + \beta_i e^{-i\omega_N t}) | \vec{\beta} e^{-i\omega_N t} \rangle \quad (6.79)$$

and to neglect it, due to the phase mismatch between the two coherent states, leading to a negligible overlap $\langle \vec{\alpha} e^{-i\omega_r t} | \vec{\beta} e^{-i\omega_N t} \rangle$. Calculating the terms proportional to

$\langle \vec{\alpha} | \vec{\xi}_i(\frac{t}{2}) | \vec{\alpha} \rangle$ and assembling all elements of (6.74) we easily arrive at

$$V_{\vec{p}\,'s',\vec{p}s} \simeq \delta_{ss'} \frac{2}{V} \frac{\alpha^2}{\omega_r} \left[\frac{Z_{eff}e^2}{(\vec{p}-\vec{p}\,')^2 + \mu_D^2} \right]^2 \cdot$$
$$\cdot \left(\frac{N}{V} \right)^2 \frac{1}{3} \frac{(\vec{p}-\vec{p}\,')^2}{2m\omega_N} \frac{\pi^2}{\omega_N^2 |\vec{p}-\vec{p}\,'|} \theta(2\omega_N - |\vec{p}-\vec{p}\,'|), \tag{6.80}$$

which yields the kernel $V(\epsilon, \epsilon')$ of the "S-wave" gap equation (see equation (6.34))

$$V(\epsilon, \epsilon') = -\lambda_{BCS} \left[1 - \left(\frac{|\epsilon - \epsilon'|}{2\omega_N v_F} \right)^3 \right] \theta \left(2\omega_N - \frac{|\epsilon - \epsilon'|}{v_F} \right) \tag{6.81a}$$

where

$$\lambda_{BCS} = \frac{2}{9} \left| \frac{\omega_N}{\omega_r} \right| \xi_{max}^2 \left[\frac{Z_{eff}e^2}{\mu_D^2} \left(\frac{N}{V} \right) \right]^2 \frac{1}{v_F} n_V, \tag{6.81b}$$

and $n_V \simeq Z_{eff}$ is the number of valence electrons.

Have we obtained SC ? Le us analyse equations (6.81) in detail. First the pairing potential is indeed attractive, implying a non-trivial solution of the gap equation (6.33). This is certainly satisfactory, but do we get decent energy cut-offs ($\simeq \omega_D$) and gaps, and most importantly the isotope effect ?

We know from the BCS analysis that the scale of the gap is determined by the cut-off energy ω_c, which in our case is (see equation (6.81a))

$$\omega_c = 2\omega_N v_F. \tag{6.82}$$

We note that according to equation (6.66) ω_N has indeed the correct isotopic mass dependence, while a look at (6.81b) shows that λ_{BCS} is mass independent due to $\left| \frac{\omega_N}{\omega_r} \right| = |1 - \dot{\phi}|^{-1}$. So we have derived a pairing potential which **does** predict the isotopic effect.

Furthermore, for a typical superconductor, like Pb, setting $Z_{eff} = 3$, $m_N \simeq 200\,\text{GeV}$ $a \simeq 2.5$ Å; $v_F \simeq 10^{-2}$, we estimate:

$$\omega_c \simeq 16 \text{ K}, \tag{6.83}$$

a rather low "Debye temperature". But let us go on and give an estimate of λ_{BCS}. Taking $\left| \frac{\omega_N}{\omega_r} \right| \simeq 10$, $\vec{\xi}_{max}^2 \simeq (0.25\text{Å})^2$, $n_V = 3$, $\mu_D = a^{-1}$, we obtain:

$$\lambda_{BCS} \simeq 0.52, \tag{6.84}$$

leading us to expect a gap $\Delta(0,0)$ of the order of ω_c, as experimentally observed. In view of the crudity of this numerical estimate any detailed discussion of experiments is certainly inappropriate at this time. However, what is gratifying about (6.83) and (6.84) is that the numbers do not come out preposterous at all but rather in the right "ball park".

We leave here our discussion of possible theoretical ideas of cold superconductivity with the feeling that what has been described appears as a very realistic and hopeful framework for explaining SC from the fundamental QED laws. I believe that a systematic effort along this line of thought will finally succeed to bring SC to the realm of scientific rationality.

Sketchy though the analysis of cold SC just presented may have been that of high T_c superconductors will be even sketchier, being simply confined to a few preliminary considerations, in search for possible QED mechanisms that in the fascinating new ceramic materials might be operative so as to raise the critical temperature T_c above 100 K.

One of the features of high-T_c superconductors that seems most remarkable is the variation of fundamental magnetic and conductance properties with the stoichiometric parameter x. What is universally found within this class of materials is that by changing x one goes from an insulating material with antiferromagnetic order (with rather high Ne'el temperature) to a conductor that has lost its antiferromagnetism but will become a superconductor at (relatively) high temperature. This is why many of the theorists, who in the last five years have devoted themselves to the great puzzles of high-T_c SC, have attached a definite significance to the loss of antiferromagnetism and the acquisition of superconductivity and viceversa with the change in x. I believe we have here a lead that should be pursued.

As we shall see in Chapter 9, the emergence of ferromagnetic phenomena appears to be fundamentally related to "strong electromagnetic coherence" involving atomic bound d-electrons that perform coherent plasma oscillations in orbits with non-zero angular moments, giving rise to ordered spin structures, ferromagnetic (spin parallel) or antiferromagnetic (spin antiparallel). Loss of such ordered structures can be understood as arising from the prevalence of quantum fluctuations, that make them only a transient phenomenon. But one of the aspects that may be important for identifying the relevant mechanism(s) for "hot" SC, is the existence of a strong coherent plasma process associated with localized d-electrons.

If this is correct we may envisage a coupling between a plasma of d-electrons with the plasma of conducting s-electrons, holes or whatever happens to conduct electricity in the particular material to be considered.

Just to get some orders of magnitude let us simply, and naively, substitute in the formulae for cold SC the plasma of nuclei with the plasma of n_d d-electrons per atom.

First the plasma frequency (6.66): in the case of the d-electrons is multiplied by a large factor, yielding cut-off frequency $\omega_c \simeq n_d \, 3 \cdot 10^3 K$. This large amplification of the cut-off frequency is quite remarkable, for in the general BCS framework ω_c sets the scale for both $\Delta(0,0)$ and T_c.

Next we derive the gap equation for a two-dimensional Fermi liquid of charge carriers, as suggested by the anisotropy of the layered structure of high T_c materials, whose planar lattice has a constant $a \simeq 3\mathring{A}$ and the interlayer distance is $c \simeq 10\mathring{A}$.

We may adapt (6.80) to our case in the form (ω_d is the d-electron plasma frequency):

$$(V_{Pair})_{\vec{p}\ 's',\vec{p}s} \simeq \delta_{ss'} \frac{2}{V} \frac{\xi_{max}^2}{\omega_r} \frac{1}{2} (\vec{p} - \vec{p}\,')^2 \left[\frac{n_d e^2}{(\vec{p} - \vec{p}\,')^2 + \mu_D^2} \left(\frac{N}{V} \right) \right]^2 . \qquad (6.85)$$

$$\cdot \frac{\pi^2}{2\omega_d^2 |\vec{p} - \vec{p}\,'|} \theta(2\omega_d - |\vec{p} - \vec{p}\,'|), \qquad (6.86)$$

and in order to obtain the kernel of the gap equation we must integrate (6.86) over the two dimensional phase space.

The kinematics of the integration (L is the length in the direction perpendicular to the CuO_2-layers, and A is the area of the layers)

$$V \int \frac{d^3\vec{q}}{(2\pi)^3} \ldots = \frac{L}{c} \int \frac{d^2\vec{q}}{(2\pi)^2} \ldots = \frac{L}{c} \frac{p_F}{v_F} A^2 \int \frac{d\epsilon}{4\pi^2} d\phi \ldots, \qquad (6.87)$$

thus we have

$$V(\epsilon, \epsilon') = \frac{\lambda}{2} \theta(2\omega_d v_F - |\epsilon - \epsilon'|) \left[\sqrt{1 - \left(\frac{\epsilon - \epsilon'}{2\omega_d v_F} \right)^2} + \left(\frac{\epsilon - \epsilon'}{2\omega_d v_F} \right)^2 \text{Ash}\left(\left(\frac{2\omega_d v_F}{\epsilon - \epsilon'} \right)^2 - 1 \right) \right]$$
$$(6.88a)$$

where

$$\lambda = \frac{1}{4c\omega_d} \left| \frac{\omega_d}{\omega_r} \right| \frac{\xi_{max}^2}{v_F} \left[\frac{n_d e^2}{\mu_D^2} \left(\frac{N}{V} \right) \right]^2 . \qquad (6.88b)$$

The most remarkable aspect of this simple formula and calculation, if we compare it with (6.81), is the appearance in the denominator of λ of the factor $\omega_d c \simeq 0.1$, which is quite large, and can be considered a bonus from the two dimensional kinematics of the layered structure.

A numerical example: taking $v_F \simeq 10^{-2}$, $n_d \simeq 3$, $\mu_D \simeq a$, $\xi_{max} \simeq 0.5\mathring{A}$, $\omega_d \simeq 20 \, eV$, we get $\lambda \simeq \frac{1}{3}$, and using the BCS formulae (6.35) and (6.36) one estimates $\Delta(0,0) \simeq 220K$ and $T_c \simeq 125K$.

In view of the crudity of this model and calculation one should not take these numbers too seriously: they only indicate that it may be fruitful to pursue these ideas further. As of the time of this writing (August 1992) this is all the "wisdom" that has been accumulated on the electrodynamical coherence mechanisms for high-T_c materials: it isn't much, but I expect it to grow considerably.

Exercises of Chapter 6

$\langle$**6.1**$\rangle$: Derive (6.11).

$\langle$**6.2**$\rangle$: Derive (6.30).

$\langle$**6.3**$\rangle$: Derive the Fröhlich potential (6.43).
    Hint: use perturbation theory with the interaction hamiltonian

$$H_I = g \int \Psi^\dagger(\vec{x},t)\Psi(\vec{x},t)A(\vec{x},t),$$

where $A(\vec{x},t)$ is a "scalar" phonon field:

$$A(\vec{x},t) = \frac{1}{\sqrt{V}}\sum_{\vec{q}}[a_{\vec{q}}e^{-i(\omega_{\vec{q}}t-\vec{q}\cdot\vec{x})} + \text{h.c.}].$$

$\langle$**6.4**$\rangle$: Derive V_{ep}, Eq. (6.44)

$\langle$**6.5**$\rangle$: Show that (6.47c) is inconsistent with the Fröhlich potential(6.44).

References to Chapter 6

1. J. Bardeen, L.N. Cooper and J.R. Schrieffer, *Phys. Rev.* **108** (1957) 1175.
2. L. N. Cooper, *Phys. Rev.* **104** (1956) 1189.
3. H. Fröhlich, *Phys. Rev.* **79** (1950) 845 .
4. E. Del Giudice, R. Mele, A. Muggia and G. Preparata, *Il Nuovo Cimento* **D15** (1993) 1279.
5. R. Balian and N.R. Werthamer, *Phys. Rev* **131** (1963) 1553 .
6. P.W. Anderson and W.F. Brinkman, *Phys. Rev. Lett.* **30** (1971) 1103 ;
   P .W. Anderson and P. Morel, *Phys. Rev* **123** (1961) 911 .

Chapter 7

JOE WEBER'S PHYSICS

7.1   The Detection of Gravitational Waves

Foremost among the very few scientists who have perceived the importance of coherence in condensed matter systems is Joe Weber, whose long scientific life begun with the discovery of nothing less than the MASER, for which surprisingly (and sadly) he was to receive no reward (and the recognition of very few people). In this Chapter we shall analyse a few of the main lines of research in which Joe Weber has been involved in the last 40 years that are all permeated with the expectations and the findings of new strange phenomena, whose explanations, as we are now going to see, rest upon the existence of the macroscopic order induced by the quantum electrodynamical coherence phenomena, that we have been discussing in this book.

In this first Section we shall be concerned with the problem of the detection of gravitational waves, which has occupied Joe Weber since his seminal work on general relativity and gravitational wave-detection that goes back to 1961[1]. The problem is well known: Einstein's General Relativity (GR) equations predict the existence of wave-solutions, the gravitational waves, which should be emitted every time a massive body develops an oscillating quadrupole mass distribution. The instances in which one can obtain large energy releases in gravitational waves are well known and include mass accretion on a black hole, member of a binary system, and supernova explosion events.

In order to produce a long awaited proof of an important prediction of GR, during the 1960's Joe Weber developed a very ingenious antenna consisting of a large suspended aluminium bar, whose exceedingly small amplitudes of deformation induced by the passage of a gravitational wave could be revealed by appropriate transducers. One can easily imagine that in presence of external disturbances and temperature effects finding the gravitational needle in the environmental and thermal stack may

145

well be a desperate enterprise. These, however, did not deter Weber, for he found means to elude both; the former by analyzing coincidence events from differently located antennas, the latter by working at the low temperatures of liquid ^{4}He. By exploiting two coincident antennas, in the United States, to which there was later added a similar antenna built by a group of Rome University, in 1968 Joe Weber produced the first evidence of coincidences that could be best (and only?) interpreted as gravitational wave's signals coming from the center of our galaxy[2]. After a first wave of enthusiasm for the important proof of a fundamental GR prediction, the scientific community reacted almost hysterically against the evidence presented by Weber, due to what was perceived as being a fundamental theoretical obstacle: an evaluation of the cross-section for the resonance absorption of the gravitational wave by the bar-antenna, due to Weber himself, which reads:

$$\sigma_{RC} = \int_{Res} \sigma(\nu)d\nu = \frac{1}{2\pi}GML^2\omega_o^{x\cdot}; \qquad (7.1)$$

where $G \simeq 4 \cdot 10^{-66} \mathrm{cm}^2$ is Newton's constant, M is the mass of the bar, L its length and ω_0 is the bar resonance frequency ($\omega_o \simeq 2\pi \cdot 10^3 \mathrm{sec}^{-1}$). Due to the smallness of Newton's constant, it is not difficult to recognize in (7.1) a very tiny cross-section, which makes the detectability of gravitational waves extremely problematic. In fact if on the basis of (7.1) one tried to evaluate how much gravitational energy should have been in the waves that induced Weber's coincidences, one would have come to the conclusion that our galaxy should have been on its way to a gravitational collapse, due to the monstrous amount of gravitational energy that was being irradiated out of its center: an obviously absurd conclusion. In the light of such acute "theoretical" impossibility, Weber's facts were dismissed as due to errors of unknown (but certainly real) origin, and the few experiments hurriedly mounted by "universally" respected scientists to "reproduce" those results, almost immediately yielded the predictable (and hoped for) "negative" results, that instantly pushed the unrepentant Weber, in spite of his brilliant achievements, to the margins of the scientific community. As we shall see the analogies with the more recent happenings of the "Cold Fusion Saga" are remarkable (and rather chilling).

As the evidence of "positive" data was accumulating in the successive years Weber, who was as troubled as everybody else by the huge discrepancy (some 6 orders of magnitude) between a reasonable gravitational wave energy irradiated out of the galactic center (where there is mounting evidence that a large black hole sits) and his coincidence data implied by (7.1), began to be convinced of the reality of his "gravitational wave" events and therefore of the complete inadequacy of (7.1) to describe the process of gravitational energy absorption of the bar antenna. He then set out to challenge the classical theory that was at the basis of (7.1), and in 1984[3] he published a calculation, based on the ideas of coherence that are the "leit-motiv" of this book. This calculation produced a very large factor $\sim N^{\frac{1}{3}}$ (N is the number of atoms of the antenna) with respect to the classical formula, that in the actual case, for which $N \simeq 10^{29}$, amounts to almost 10 orders of magnitude, much too good! Even though I do not agree with the actual details of the calculation, what is certainly of fundamental importance in it is the clear idea that the simple classical formula cannot be anywhere near the correct description of the cross-section, for it misses completely the point that the exchange of energy between the gravitational wave —the gravitons— and the antenna takes place with the elementary constituents (the atoms) being in the highly correlated state that, as we have seen in Chapter 5, characterizes the solid state.

The subsequent coincidence data gathered in 1987 during the explosion of the supernova SN 1987A further corroborated Weber's expectation that (7.1) should be some 5 or 6 orders of magnitude smaller than the real cross-section[4].

In order to evaluate such cross-section let us consider the bar antenna as an ensemble of ($N \simeq 10^{29}$) atoms performing collective oscillations with typical frequency $\omega_r (\simeq 10^3 \mathrm{cm}^{-1})$ described by a collective wave-function of the type (5.3.5). We shall see that it is the collective character of the coherent motion that leads to the striking amplification over eq. (7.1) necessary to fit the data.

Let us begin with some normalizations. We consider a monochromatic gravitational wave moving in the 3-direction with a given polarization, say $+$, and pulsation ω; its amplitude is (we obviously neglect the space-dependence)

$$A_+(\omega, t) = A_+(\omega)e^{-i\omega t} + A_+(\omega)^* e^{i\omega t}. \tag{7.2}$$

The bar antenna is a cylinder with axis along the 1-direction; the relevant components of the energy-momentum tensor are

$$T_{00} = -T_{03} = \frac{1}{8\pi G}\omega^2 |A_+(\omega)|^2 \tag{7.3}$$

In the language of second quantization, introducing the "ether oscillators" of the gravitational field we have:

$$T_{00} = \frac{\omega}{V}a_+^\dagger(\omega)a_+(\omega) = \frac{\omega^2}{8\pi G}|A_+(\omega)|^2 \tag{7.4}$$

where $a_+(\omega)(a_+^\dagger(\omega))$ is the annihilation (creation) operator of the mode with frequency ω. Eq. (7.4) allows us to establish the connection:

$$A_+(\omega) = \left(\frac{8\pi G}{\omega V}\right)^{\frac{1}{2}} a_+(\omega), \tag{7.5}$$

between the classical gravitational field amplitude $A_+(\omega)$ and the coherent amplitude of the operator $a_+(\omega)$. Calling m the mass of the generic atom and $\vec{x}$ its position, the interaction Hamiltonian atom-gravitational field is[5]

$$H_I = \frac{m\omega^2}{4}[A_+(\omega)e^{-i\omega t} + A_+(\omega)^*e^{i\omega t}](x_1^2 - x_2^2) \qquad (7.6)$$

and, using first order perturbation theory, we may calculate the transition amplitude of the bar:

$$
\begin{aligned}
M(\omega, t)_{\vec{\alpha} \to \vec{\beta}} &\simeq -i \int_0^t dt' \frac{m\omega^2}{4}[A_+(\omega)e^{-i\omega t} + A_+(\omega)^*e^{i\omega t}]\langle\vec{\beta}|(x_1 + \xi_1)^2|\vec{\alpha}\rangle \\
&\simeq -i \int_0^t dt' \frac{m\omega^2}{4}[A_+(\omega)e^{-i\omega t} + A_+(\omega)^*e^{i\omega t}]\langle 2x_1\rangle\langle\vec{\beta}|\xi_1(t)|\vec{\alpha}\rangle,
\end{aligned}
\qquad (7.7)
$$

where $\langle 2x_1\rangle$ is the sum over the atomic equilibrium position over half of the bar and is equal to $\frac{LN}{4}$, L being the length of the bar. The reason why we separate the two halves of the bar is that we are studying the absorption of the classical gravitational field in a non-resonant situation over a time $\tau \simeq \frac{2\pi}{\omega_r}$ that, as we have seen in Section (3.7), corresponds to the relaxation time of an excited state of the bar to the Coherent Ground State (CGS), through emission of photons and phonons. The interaction Hamiltonian (7.6), which represents the energy exchange between the matter system and the gravitational wave and oscillates in time and space, shows that absorption from the bar can only take place during those periods of oscillation (of the matter oscillators) in which the expectation value of H_I is positive, and this happens in two different half cycles for the two halves of the bar. Another relevant observation is that the absorption of gravitational energy takes place during a time τ much smaller than the period of oscillation of the gravitational wave $T_g = 2\pi/\omega$, for the system wave-antenna is "isolated" —and unitary— only during a time interval shorter than the typical periods τ of the internal coherent dynamics: over time intervals longer than τ the energy exchanges with the gravitational wave are in fact absolutely negligible with respect to the energy exchanges inside the antenna. However it should be noted that such exchanges, following the absorption of gravitational energy, are modulated by the pulsation ω of the wave, and when we observe the dynamics of the bar over long periods of time this energy will have a spectral distribution that is identical to the spectral distribution of the gravitational energy in the wave.

For the rate of absorption we may thus write:

$$R_{abs} = \frac{2}{\tau}\sum_{\vec{\beta}}|M(\omega, \tau)_{\vec{\alpha} \to \vec{\beta}}|^2, \qquad (7.8)$$

where $\tau \simeq 2\pi\omega_r$, and the factor of two stems from the identical contribution of the two halves of the bar. Evaluating (7.8) we obtain:

$$
\begin{aligned}
R_{abs} &\simeq \frac{\omega_r}{\pi}\frac{(m\omega)^2}{16}\frac{(LN)^2}{16}4|A_+|^2\int_0^\tau dT \int_{-2T}^{2T}\langle\vec{\alpha}|\xi_1(t)\xi_1(0)|\vec{\alpha}\rangle dt \\
&= \frac{\omega_r}{2\pi}\frac{(m\omega^2)^2}{64}(LN)^2|A_+|^2\frac{|\vec{\xi}_{max}|^2}{3}\frac{1-\cos^2\omega_r\tau}{\omega_r^2} \\
&\simeq \frac{ML^2N}{2\pi\omega_r}\frac{m\omega^4}{192}|\vec{\xi}_{max}|^2|A_+(\omega)|^2,
\end{aligned}
$$
(7.9)

where we have averaged over times of absorption τ of the order of $(\frac{2\pi}{\omega_r})$, and $\vec{\xi}_{max}$ (see Section (5.3)) is the maximum oscillation amplitude of the Al-nuclei ($|\vec{\xi}| \simeq 0.25\text{Å}$). In order to obtain the cross-section we must divide R_{abs} by the graviton flux $\frac{T_{oo}}{\omega}$, which according to (7.4) is equal to $\frac{\omega}{8\pi G}|A_+(\omega)|^2$, thus we have for the coherent cross-section[a]

$$
\sigma_c = \frac{8\pi G R_{abs}}{\omega|A_+(\omega)|^2} = \frac{GML^2\omega^2}{2\pi}\frac{8\pi m}{192}\left(\frac{\omega}{\omega_r}\right)|\vec{\xi}_{max}|^2N.
$$
(7.10)

Integrating over the frequency interval $\Delta\omega$ around the frequency ω of the resonant bar mode we obtain the cross-section amplification factor λ_c:

$$
\lambda_c = \frac{\sigma_c}{\sigma_{CR}} = \left(\frac{\Delta\omega}{\omega}\right)\left[\frac{m\omega_r|\vec{\xi}_{max}|^2}{4g}\right]\left(\frac{\omega}{\omega_r}\right)^2N.
$$
(7.11)

Putting numbers in: $\left(\frac{\Delta\omega}{\omega}\right) \simeq 10^{-3}, \omega_r \simeq 10^3\text{cm}^{-1}, |\vec{\xi}_{max}| \simeq 0.25\text{Å}, m = 1.5\cdot 10^{15}\text{cm}^{-1}, \omega \simeq 2\cdot 10^{-7}\text{cm}^{-1}, N \simeq 10^{29}$ we get:

$$
\lambda_c \simeq 10^6,
$$
(7.12)

which in view of the approximate nature of the calculation, and of our neglect of the reduction of the coherent fluid density due to thermal fluctuations, is a rather remarkable result.

The great theoretical obstacle towards accepting that the large number of observations by J. Weber (and later also by the Rome group) of coincidences in the bar antenna are due to gravitational waves, is thus seen to be handsomely overcome by the simple idea, discussed in Section (5.3), that the solid state finds its *raison d'être* in the occurrence of coherent oscillations of the atomic nuclei. We have just seen that the energy exchange of a gravitational wave with a matter system in this kind of state,

---

[a]A related derivation is to be found in Ref. [6]

as opposed to a gas of incoherent matter points, is greatly amplified by a factor λ_c of the order of 10^6. Such large factor, that accordingly decreases the estimates of the energy content of the detected gravitational waves, brings all observations over the last fifteen years back to the domain of scientific rationality. It goes without saying that all this implies that a new important "eye" on the secrets of our Universe has just been open.

7.2   The Detection of Neutrinos

The detection of gravitational waves is not the only "impossible" challenge that Weber took up, and brought to a successful conclusion: there is another particle that is almost as elusive as the graviton, the neutrino. The neutrino interacts so weakly with matter that it took almost 30 years to show that the theoretical proposal by Pauli, cleverly conceived to avoid giving up the principle of conservation of energy in β-decay, was indeed realized in nature in the form of (almost) massless particles, carrying spin $\frac{1}{2}$ and no charge, called by Fermi "neutrinos". Modern particle physics has demonstrated that there exist three types of neutrinos (ν_e, ν_μ, ν_τ) that are coupled to the intermediate vector bosons ($W^\pm, Z^\circ$) of the electro-weak interactions only through their left-handed components. The detection of neutrinos in high energy physics involves processes of the "inverse-β-decay" type:

$$\nu_{e,\mu} + N \rightarrow e^-(\mu^-) + \text{anything} \qquad (7.13)$$

where N is a nucleon of a given target, and the final state may be a single nucleon (elastic scattering) or a highly excited hadronic state (deep inelastic scattering). The experimental apparatus necessary to reveal the charged lepton originating from ν-scattering is typically a huge array of chambers and counters that costs many tens of millions of US $, and can only be installed in the large high energy physics laboratories. Weber's idea on the other hand was again to put to work the subtle, but robust coherence of the nuclei of a highly stiff crystal, to greatly increase (by the usual factor of N implied by coherence) the scattering cross-section. In this way one could detect neutrinos with table-top experiments of very low cost and high flexibility.

The process that we shall analyse here is the elastic scattering of electron-neutrinos of momentum $\vec{p}$ (along the 3-direction) off the nuclei of a stiff crystal, like sapphire or silicon,

$$\nu_e(\vec{p}) + \text{crystal} \rightarrow \nu_e(\vec{p}) + \text{crystal} \qquad (7.14)$$

The (effective) interaction Hamiltonian is the 4-fermion Hamiltonian describing the neutral-current interaction:

$$H_I = \frac{G_F \cos^4 \theta_w}{\sqrt{2}} \int d^3 x \nu(x)^\dagger (1-\gamma_5) \nu(x) \psi_N^\dagger(\vec{x}, \vec{\xi}, t)(\tau_3 - 4Q \sin^2 \theta_w) \psi_N(\vec{x}, \vec{\xi}, t), \quad (7.15)$$

where $G_F \simeq 10^{-5} m_p^{-2}$ is the Fermi constant, θ_w is the weak interaction angle ($\sin^2 \theta_w \simeq \frac{1}{4}$), τ_3 is the third component (multiplied by 2) of the isospin operator, and Q the charge. By factoring out in (7.15) the Dirac spinors we may write:

$$H_I = \frac{G_F \cos^4 \theta_w}{V\sqrt{2}} \sum_{\vec{p}'} u(\vec{p}')^\dagger (1 - \gamma_5) u(\vec{p}) e^{-i(|\vec{p}|-|\vec{p}'|)t} \int_{\vec{x}, \vec{\xi}} d^3 x e^{i(\vec{p}-\vec{p}')\vec{x}}$$

$$(N_p(1 - 4\cos^2 \theta_w) - N_n) \psi_N^\dagger(\vec{x}, \vec{\xi}, t) \psi_N(\vec{x}, \vec{\xi}, t), \quad (7.16)$$

where N_p (N_n) is the number of protons (neutrons) in the nuclei. Using first order perturbation theory for the Feynman amplitude for elastic scattering we have:

$$M(\vec{p} \to \vec{p}') = -i \int_{-\infty}^{\infty} dt \langle \vec{\alpha} | H_I(t) | \vec{\alpha} \rangle$$

$$= \frac{-iG \cos^4 \theta_w}{\sqrt{2}} (2\pi) \delta(|\vec{p}| - |\vec{p}'|)(2Z - A) u(\vec{p}')^\dagger \gamma_0 (1 - \gamma_5) u(\vec{p}) F(\vec{p} - \vec{p}'), \quad (7.17)$$

where according to (5.88) the form factor $F(\vec{p} - \vec{p}')$ is given by:

$$F(\vec{p} - \vec{p}') = N \exp\left\{ -\left[\frac{(\vec{p} - \vec{p}')^2}{2m\omega_N} - i\frac{\vec{q}(\vec{\alpha} + \vec{\alpha}^*)}{4\sqrt{2m\omega_N}} \right] \right\}. \quad (7.18)$$

Recalling the discussion in Section (5.3), the factor N, expressing the coherence of the momentum transfer process, is most remarkable for it greatly amplifies the minute incoherent neutrino cross-section. Following the well known rules we can now compute the cross-section in a straightforward fashion, and we get ($\cos \theta = \frac{\vec{p} \cdot \vec{p}'}{pp'}$, $N_{neutr} = N_p(1 - 4\sin^2 \theta) - N_n$)

$$\frac{d\sigma_{elastic}}{d\cos\theta} = \frac{(NG_F p)^2}{4\pi} (\cos^4 \theta_w N_{neutr})^2 (2 - \cos\theta) e^{-\frac{p^2(1-\cos\theta)}{m\omega_N}}, \quad (7.19)$$

and integrating over the angle we have:

$$\sigma_{elastic} \simeq \frac{(NG_F)^2}{a\pi} (\cos^4 \theta_w N_{neutr})^2 m\omega_N \left(1 + \frac{m\omega}{p^2}\right) \quad (7.20)$$

which is essentially J. Weber's result[7]. In order to appreciate the implications of (7.20), let us set $N \simeq 10^{25}$ (1 Kg of crystal), $m \simeq 2.5 \cdot 10^{15} \text{cm}^{-1}, \omega_N \simeq 10^3 \text{cm}^{-1}, p \simeq$

1MeV, we get a cross-section:

$$\sigma_{elastic} \simeq 3 \cdot 10^2 \left(\cos^4 \theta_w N_{neutr} \right)^2 \text{cm}^2 \tag{7.21}$$

a result which is so big that it violates unitarity, showing the necessity of a more refined calculation. However the message from this simple evaluation is quite clear: the probability that a neutrino is absorbed in a stiff crystal is close to 1. But in order to be detected the incoming neutrino must transfer momentum to the crystal, and the average momentum transferred is easily calculated (p is the incident momentum)

$$\langle \Delta p \rangle \simeq p \left(\frac{m \omega_N}{p^2} \right); \tag{7.22}$$

for $p \simeq 1$MeV, one estimates $\Delta p \simeq 10$KeV. Taking a ν-flux of 10^{12} neutrinos per second one estimates an overall force on the crystal of about 10^{-6} dyne, a force that can be measured with a torsion balance, as Weber has brilliantly shown. Note that the stiffer the crystal, the larger $m\omega_N$, the bigger $\langle \Delta p \rangle$, the easier the detectability of the neutrino beam. As Weber reminds us[7], several hundred experiments confirm the large cross-section (7.20).

To briefly summaries this Section: in spite of the smallness of the Fermi constant $G_F \simeq 4 \cdot 10^{-33}$cm^2 the huge enhancement of $\sigma_{elastic}$ (Eq. (7.20)) by the factor $N \simeq 10^{25}$ makes it possible for a typical neutrino beam to exert on a stiff crystal (large ω_N) a measurable force of the order of 10^{-6} dyne. Without such amplification, due to coherent scattering, the mean free path of a neutrino in an incoherent medium would be about 10^{18}cm, $i.e.$ one light year! Weber was not only able to understand that a coherent medium would completely do away with the elusiveness of the neutrino, but he could demonstrate this with simple and elegant experimental means.

References to Chapter 7

1. J. Weber, *General Relativity and Gravitational Waves*, Iterscience (John Wiley, New York-London 1961).
2. J. Weber, *Phys. Rev. Lett.* **20** (1968) 1307.
3. J. Weber, *Found. Phys.* **14** (1984) 1185.
4. M. Aglietta et al., *Il Nuovo Cimento* **12C** (1989) 75.
5. C.W. Misner, K.S. Thorn and J.A. Wheeler, *Gravitation* (Freeman and Co. San Francisco, 1973).
6. G. Preparata, *Mod. Phys. Lett.* **5A** (1990) 1.
7. J. Weber, *Phys. Rev.* **31C** (1985) 1468.

Chapter 8

TOWARDS A THEORY OF COLD FUSION PHENOMENA[a]

8.1 The "impossible" phenomenology of Cold Fusion (CF)

On March 23rd 1989 a strange, unbelievable announcement struck the scientific community: the nuclear fusion processes that are deemed to fuel the stars, at the enormous temperatures and pressures that exist in their interiors, could be induced at room temperature and atmospheric pressure in an earthly laboratory. Martin Fleischmann and Stanley Pons, two chemists then working at the University of Utah in Salt Lake City, announced in a heated press conference that a series of experiments conducted in the previous years had given them the evidence that a large quantity of energy, in the form of heat, could be produced in the electrolysis of heavy water (D_2O) with a palladium (Pd) cathode, that could not be of chemical origin (due to the magnitude of the effect). And if not of chemical origin, the two chemists concluded, the **most conservative** hypothesis was that the observed excess heat might be due to deuterons pairs fusing within the Pd metal matrix[1].

The "miracle" that once inside the Pd lattice the highly repulsive Coulomb barrier could be overcome so as to yield the large overlap probabilities of the deuterons at nuclear distances (a few Fermis, 1 F $= 10^{-13}$ cm) necessary to trigger the fusion process, was however not the only one occurring in the strange phenomena that Fleischmann and Pons had collected in years of experimentation. Even admitting, in fact, that for some unexpected reason peculiar electronic distributions inside Pd screen considerably the D-D Coulomb repulsion, other more spectacular "miracles"

---

[a]Since this Chapter was written (August 1992) some theoretical developments within QED coherence in matter have occurred that in part supersede some of the analysis carried out in this Chapter, they are reported in Ref. [11]. However, I have decided (September 1994) not to rewrite this Chapter, in order to keep a testimony of the difficult and minding road the research program described in this book must go through to shed light upon very mysterious phenomena, such as those of Cold Fusion.

emerge from the log-books of the Utah electrochemists. Indeed the large excess heat is seen to have no correlation whatsoever with the two leading reaction channels:

$$D + D \rightarrow \ p \ (3.02 \, \text{MeV}) +^3 H \ (1.01 \, \text{MeV}) \ , \tag{8.1}$$

$$D + D \rightarrow \ N \ (2.45 \, \text{MeV}) +^3 He \ (0.82 \, \text{MeV}) \ , \tag{8.2}$$

that are known to occur with about the same probability when two deuterons fuse (at high energy, or temperature) *in vacuo*. When looking for either neutrons or ^{3}H (tritium) Fleischmann and Pons (FP) found very little traces of these particles[1], the mismatch between the nuclear "ashes" (8.1) and (8.2) being of the "astronomical" order of about 10^9. In other words, the energy collected in the form of heat during the electrolysis of D_2O could not be due to the dominant DD-fusion processes *in vacuo*, but it should involve some other unknown dynamical mechanism. If not (8.1) and (8.2), the burning questions now are, what are the "ashes" and, most important, why does this happen? FP conjectured the process involved to be

$$D + D \rightarrow \ ^4He + heat \tag{8.3}$$

where the excess energy — 23.4 MeV — of the final state should be somehow released to the lattice, which would then relax it in the form of the observed heat. Later on yet another "miracle" appeared to occur in the already "impossible" CF phenomenology: measurements of tritium during electrolysis[2] or in D_2 absorption in Titanium[3] showed that the branching ratio:

$$B \left(\frac{D + D \rightarrow p +^3 H}{D + D \rightarrow N +^3 He} \right) \simeq \ 10^6 \div 10^8 \ , \tag{8.4}$$

and not about 1, as observed in standard (*in vacuo*) nuclear experiments.

All such "miracles", whose destructive potential for generally accepted theories and scientific strategies was keenly perceived by the scientific community, created an immediate wave of skepticism especially among the physicists. The difficult reproducibility of the FP experiments by scientists and amateurs alike, who had not fully comprehended many subtle aspects of the experimental protocols necessary to first load the Pd-cathode with deuterons and then induce the excess heat producing reactions, converted this wave of skepticism into a vicious full scale campaign, both in the media and in the scientific circuits, to discredit, worse to criminalize FP and everybody dared to take a positive or an open attitude towards the "impossible" phenomenology. Surprisingly, the negative attitude has been so strong and so tenacious that has survived the accumulation of a remarkable number of confirmations and developments of the original FP phenomenology. At the time of this writing (August 1992) the field is "coming of age" and the 3rd Annual Conference of Nagoya (Japan) is expected to further establish the true scientific nature of Cold Fusion phenomena.

To summarize the above brief sketch of CF phenomena that have been reported so far[b]:

(i) during the electrolysis of heavy water with a Pd-cathode many different groups have reported the production of excess heat over extended periods of time (weeks), amounting to several W/cm^3 of Pd;

(ii) in a similar type of experiments, with cathodes made of an Ag-Pd alloy, FP have observed powers up to a few KW/cm^3 [c];

(iii) in situ measurements of the loading ratio $x = \dfrac{D}{Pd}$ have shown that excess heat is produced only when the "bulk" x exceeds a value x_0, higher than the β-phase loading ratio $x_\beta \simeq 0.6$;[d]

(iv) there are indications that the amount of ^{4}He present in the evolution gases of the electrolysis compares with the excess heat, lending credence to the dominance of the reaction channel (8.3)[e];

(v) several experiments report substantial tritium production during FP-type experiments; the amount of tritium found is however four to five orders of magnitude short to account for the excess heat production;

(vi) 2.45 MeV neutrons emission has been reported in FP-type electrolysis[5] as well as in experiments where titanium is loaded with deuterons from the gas phase[6]. The typical rates of neutron emission are at least nine orders of magnitude lower than it would be necessary to "explain" the excess heat in terms of the conventional DD-fusion reactions (8.1) and (8.2);

(vii) in absence of well defined, controlled conditions, the reproducibility of CF-phenomena is totally erratic, explaining the vast number of early unsuccessful trials.

In the light of this phenomenology, which has been consistently replicated in the three years passed since the first announcement, the exacting demands that are imposed upon any attempts to **explain** CF can be reduced to the following questions:

[b]For a panoramic of Cold Fusion phenomenology as of 1991, see Ref. [4].
[c]Cfr. S. Pons' contribution in Ref. [4]
[d]Cfr. M. McKubre's contribution in Ref. [4]
[e]Cfr. M.H. Miles' contribution in Ref. [4]

(a) how can it happen that the Coulomb barrier between two fusing deuterons can be screened so effectively as to yield overlap probabilities some forty orders of magnitude higher than expected? (Miracle # 1)

(b) how can it happen that excess heat is essentially uncorrelated with either neutrons or tritium production? (Miracle # 2)

(c) how can it happen that the two usual channels of DD-fusion have a branching ratio so different from the well known one? (Miracle # 3)

We shall see in the next Sections that all such "miracles" can be comfortably accommodated, like the other miracles discussed in the preceding Chapters, in the general framework of the QED coherence of matter.

8.2 Strange phenomena occurring at the surface of palladium

The remarkable "affinity" for hydrogen (and its isotopes, deuterium and tritium) of metals such as Pd and Ti is known since a long time, but to this day the deep fundamental reason why, for instance, a piece of Pd absorbs from the gas phase H_2-molecules, incorporating them in the lattice in the form of protons (and its isotopes D and T) in large quantities (up to $x_\beta = \frac{D}{Pd} \simeq 0.6$), seems to be thoroughly eluding us. As discussed in Ref. [7], there is a series of perhaps minor "miracles" that must happen at the surface of Pd in order to convert the initial H_2 (or D_2) molecule into the final configurations, comprising the **independent** nuclear (proton and deuteron) and electronic components. And these "miracles" cannot be explained in terms of few-body short-range electrostatic forces, due to their inability to produce the very strong electric fields needed to overcome an energy barrier about 30 eV high (such is the energy that must be fed to the H_2-molecule in order to separate it into its elementary components). Can we understand this as due to QED coherence?

Let us observe that at the surface of a metal, such as Pd, there are "evanescent-wave" configurations of the coherent e.m. fields associated with both electrons' and ions' coherent plasma oscillations, discussed in Chapter 5.

Such configurations are very similar to the e.m. field configurations that form in the vicinity of total reflection mirrors, and can be described as follows: let the interface I separate space in two semispaces ($z \gtrless 0$) and assume that on the interface I the "classical" e.m. field $\vec{A}(\vec{x}, t)$, associated with the relevant plasma, is constructed with wave-numbers $|\vec{k}| = \omega_p$ that lie in the plane $x - y$. This assumption is quite natural, for it is reasonable that the e.m. modes with momentum in the z-direction had been radiated away at the inception of the coherent process (during the run-away). Writing the free e.m. field equation,

$$\Box \vec{A} = 0, \tag{8.5}$$

valid in the semispace $z > 0$, and looking for solutions of the type:

$$\vec{A}(\vec{x}, t) = \left[e^{-i\omega_r t} \vec{A}(x, y, z) + c.c. \right], \tag{8.6}$$

where $\omega_r = \omega_p(1 - \dot{\phi})$ (see Eq. (5.46)), with the boundary condition ($\vec{k}^2 = \omega_p^2$):

$$\left(\frac{\partial^2}{\partial x^2} + \frac{\partial^2}{\partial y^2} \right) \vec{A}(x, y, z) = -\vec{k}^2 \vec{A}(x, y, z), \tag{8.7}$$

we readily derive from (8.5)

$$\vec{A}(x, y, z) = \vec{A}(x, y, 0)\, e^{-\lambda z}, \tag{8.8a}$$

with

$$\lambda = \sqrt{\omega_p^2 - \omega_r^2} \geq 0. \tag{8.8b}$$

The solution (8.8) of Eq. (8.5) is the sought "evanescent wave" configuration, that describes how the coherent e.m. field, associated with the relevant plasma oscillations, extends outside the metal. How is such evanescent wave going to affect a H_2-molecule that happens to land on the surface of Pd? Let us drastically simplify our problem by reducing the configuration space of the D_2-molecule to a two-dimensional space comprising the ground state $|0\rangle$ and the "fully ionized state" $|i\rangle$, whose energy difference $\omega_i \simeq 30\,\text{eV}$. The interaction Hamiltonian can be written:

$$H_{int} = -\vec{E} \cdot \vec{D}, \tag{8.9}$$

where $\vec{D}$ is the electric-dipole operator connecting the two states $|0\rangle$ and $|i\rangle$, and $\vec{E}$ is the electric field associated with the relevant coherent plasma (A is, as usual, the coherent e.m. amplitude):

$$\vec{E} \simeq \vec{u}\, \sin(\omega_r t) \left(\frac{2N\omega_r^2}{V\omega_p} \right)^{1/2} \left(\frac{2\pi}{3} \right)^{1/2} \cdot A, \tag{8.10}$$

as can be readily derived from the developments of Section (5.1).

The solution of our problem requires the solution of the Schrödinger equation:

$$i\frac{\partial}{\partial t}|\chi\rangle = \left[\frac{\omega_i}{2}\sigma_3 + \frac{\omega^*}{2}\left(e^{-i\omega_r t}\sigma_+ + e^{i\omega_r t}\sigma_- \right) \right] |\chi\rangle \tag{8.11}$$

where, as usual, $\sigma_{\pm} = \frac{1}{2}(\sigma_1 \pm i\sigma_2)$, and

$$\omega^* = \left(\frac{N\omega_r^2}{V\omega_p}\right)^{1/2} \left(\frac{2\pi}{3}\right)^{1/2} Ae|\xi_{0i}| \tag{8.12}$$

$e|\xi_{0i}|$ being the dipole transition matrix element between the ground state and the ionized state. From the complete solution of (8.11) [Ex. ⟨8.1⟩], substituting the appropriate reasonable values ($\omega_i \simeq 30\,\text{eV}$, $\omega_r \simeq 0.2\omega_p$, $\omega_p \simeq 10\,\text{eV}$, $|\xi_{0i}| \simeq 1\,\text{Å}$, $N/V \simeq 6.5 \cdot 10^{22}\,\text{cm}^{-3}$), for the ionization probability one obtains:

$$P_i^{(t)} = \frac{\omega^{*2}}{(\omega_i - \omega_r)^2 + \omega^{*2}} \sin^2 \sqrt{(\omega_i - \omega_r)^2 + \omega^{*2}} \frac{t}{2}, \tag{8.13}$$

which, in view of $\omega^* \simeq 10\,\text{eV}$, shows that over times longer than $\dfrac{\pi}{\omega_r - \omega_i} \simeq 10^{-16}\,\text{sec}$ the probability that a H_2-molecule be ionized on the surface of Pd is of the order of 5%: a non-negligible probability.

Does the coherent field work in order to ionize the molecule that arrives at its surface? The answer, that can be easily obtained by computing the time average of H_{int} over the solution of the Schrödinger equation, is negative: the plasma spends no energy to crash the H_2-molecule and offer it in the fully ionized state to the metal lattice. This unexpected result, that defies our intuition based on the general idea that any "activation process" (such as, in our case, ionization) must require a well defined energy exchange between the activation agent and the activated system, shows that in presence of a "strong" coherent time-dependent e.m. field an atomic or molecular species can be dynamically activated, without a net (over long periods of time) energy exchange. Such activation simply corresponds, within the CD domains of the e.m. field, to a **rearrangement** of the bound *in vacuo* configuration into an unbound one which, differently from what happens *in vacuo* (or equivalently in the PGS), does not require surmounting any energy barrier. People familiar with the notion and the facts of catalytic behaviour may recognize in the dynamical mechanism we have just analysed a very good candidate for a satisfactory general scientific explanation of catalysis.

But there is at least another strange phenomenon that occurs at the Pd-surface during FP electrolysis, namely the formation of a double layer whose charge density is equal to the proton (deuterium) concentration. The presence of this double layer proves essential for reaching through electrolysis the high loading ratios x that are observed to be associated with the production of excess heat. How can we understand the formation of this double layer? By what physical mechanism(s) the Pd-surface is able to displace the electrons, which neutralize the D-ions, further away from it than the deuterons? The dynamics of both electrons and deuterons on the Pd-surface is governed by the Hamiltonian:

$$H = \frac{(\vec{p} + e\vec{A})^2}{2m} + V_S \, , \tag{8.14}$$

where $\vec{A}$ is the coherent e.m. field existing in the form of the "evanescent wave" configuration discussed above, and V_S is the static potential binding the particles to the Pd-surface. In spite of its strongly oscillating character (with pulsation ω_r) the coherent e.m. field generates a non-zero static "ponderomotive force",

$$V_{pm} = \frac{e^2 \langle \vec{A}^2 \rangle}{2m} \, , \tag{8.15}$$

due to the non-vanishing time-average $\langle \vec{A}^2 \rangle$. The mass-dependence of V_{pm} implies that its strength for electrons is about 4000 times larger than for deuterons. Furthermore the direction of the ponderomotive force $\vec{F}_{pm} = -\vec{\nabla} V_{pm}$ is clearly along the normal to the surface, directed towards the outside. From the developments of Chapter 5, considering the plasma of Pd-nuclei (see Section (8.3)) we may write:

$$\langle \vec{A}^2 \rangle = \frac{2g^2}{\omega_N} \alpha_{max}^2 \left(\frac{N}{V} \right) e^{-2\lambda z} \, , \tag{8.16}$$

where ω_N is the plasma frequency of the Pd-nuclei, $\alpha_{max}^2 \simeq (m\omega_N)\xi^2$ is the maximum amplitude of the plasma oscillations ($\xi \simeq 0.1\mathring{A}$), and λ is given by (8.8b). Putting in numbers, we get for the ponderomotive force on the electrons:

$$(F_{pm})_z = -\frac{\partial V_{pm}}{\partial z} \simeq \frac{e^2}{2m_e} 4g^2 \left(\frac{N}{V} \right) \alpha_{max}^2 \simeq 5 \cdot 10^8 \, \text{eV/cm}, \tag{8.17}$$

that, modelling V_S as a harmonic oscillator potential $V_S \simeq V_0(\frac{z}{a})^2$, with $V_0 \simeq 5\,\text{eV}$ and $a \simeq 1\mathring{A}$, shifts the electrons' equilibrium position from $z = 0$ by about 0.7 $\mathring{A}$. The double layer's mystery thus seems to disappear.

8.3   The plasmas of CF[f]

In order to address the crucial issue of the nuclear fusion reaction involving the deuterons that pack the Pd-lattice we must have a rather detailed understanding of

---

[f]This Section contains an early stage of the development of the dynamical description of the plasmas of CF. I have decided not to rewrite this Section for two reasons: first to give flavour of the state of rapid development of the ideas of QED coherence and second to give a hopefully live picture of how these ideas have evolved into a rational picture of CF. For a much improved treatment see Ref. [8].

the environment in which such nuclear processes will eventually take place[8].

As argued in Chapter 5, all the important "coherent" actors of the CF "play" are organized in plasmas, characterized by a plasma frequency ω_p and a maximum oscillator amplitude α_{max}. Let us now review the different relevant plasmas, starting with those that are already there before the deuterons and their neutralizing electrons enter the Pd-lattice.

Let us begin with the d-**electrons plasma**. In the ideal case the plasma frequency is given by ($N/V \simeq 6.5 \; 10^{22} \, \text{cm}^{-3}$ is the Pd-density, $n_d{=}10$):

$$\omega_{de} = \frac{e}{m_e^{1/2}} \left(n_d \frac{N}{V} \right)^{1/2} = 30 \, \text{eV}. \tag{8.18}$$

In the real case, as we have argued in Section (5.2), this value gets enhanced by a factor about 1.38, due to the unscreened distribution of the neutralizing charged fluid (the Pd-ions, in our case). Thus, as a rather reliable estimate, we may set for the d-electrons' plasma frequency:

$$\omega'_{de} \simeq 41.4 \, \text{eV}, \tag{8.19a}$$

to which there corresponds a CD size

$$\lambda_{de} = \frac{2\pi}{\omega_{de}} \simeq 300 \text{\AA} . \tag{8.19b}$$

The maximum amplitude of the plasma oscillations is given by:

$$|\vec{\xi}_{max}| = (m_e \omega'_{de})^{-1/2} |\alpha_{max}| \simeq 0.61 |\alpha_{max}| \, \text{\AA} , \tag{8.20}$$

and assuming $|\alpha_{max}| = 1$, which is quite natural if we think of the d-electrons plasma as a rather localized plasma[g], we conclude that the d-electrons' plasma oscillates in a (arbitrary) direction $\vec{u}$ with maximum excursion about 0.6Å. From the results of Section (5.2) the coherent electromagnetic field associated with this plasma is (neglecting any space-dependence)[h]

$$A_k^{de} = \frac{1}{2} \left(\frac{n_d N}{2\omega_{de} V} \right)^{1/2} \left(\frac{8\pi}{3} \right)^{1/2} A(u_k e^{-i\omega_r t} + \text{c.c.}) \tag{8.21}$$

---

[g]Note that in any case $|\alpha_{max}| \geq 1$, the equality sign holding in the two-level plasma approximation.
[h]The factor $1/2$ is noteworthy, for it stems from the fact that the $\vec{k}$-modes contributing to the "evanescent wave" are one half of all the modes (they belong to the circle of the plane of the interface)

where $A \simeq 0.8$ and $\omega_r = \omega_{de}(1 - \dot{\phi}) = -10.4\,\mathrm{eV}$. The ponderomotive force acting on a single electron due to this plasma is

$$F_{pm} = -\frac{\partial}{\partial z}V_{pm} \simeq \frac{e^2 n_d N}{2 V m_e}\left(\frac{8\pi}{3}\right) A^2 = 2.5 \cdot 10^8\,\mathrm{eV/\,cm}, \qquad (8.22)$$

of the size of (8.17).

Getting now to the ions' plasma, we may visualize it as comprising the Pd-nuclei surrounded by the deeply bound electrons, carrying thus a positive charge $Q = n_d e$. The ideal plasma frequency is:

$$\omega_N = \frac{n_d e}{m_{Pd}^{1/2}}\left(\frac{N}{V}\right)^{1/2} \simeq 0.2\,\mathrm{eV}, \qquad (8.23)$$

from which, by the usual argument, we get the more realistic estimate:

$$\omega_N' = 1.38\,\omega_N = 0.28\,\mathrm{eV}, \qquad (8.24a)$$

yielding a CD size:

$$\lambda_N = \frac{2\pi}{\omega_N'} \simeq 4.5 \cdot 10^{-4}\,\mathrm{cm}. \qquad (8.24b)$$

The maximum amplitude for plasma oscillations $|\vec{\xi}_{max}|$ can be estimated to be about 0.1 Å, corresponding to the distance of the deeply bound s-electrons from the nucleus. From this we can also estimate the maximum oscillator amplitude:

$$|\alpha_{N,max}| \simeq .1\,\text{Å} \qquad (m_N \omega_N')^{1/2} \simeq 8.6 , \qquad (8.25)$$

from which we derive a coherent e.m. field amplitude (see Eq.(5.1.42c))

$$A_k^N = \frac{1}{2}\left(\frac{N}{2V\omega_N'}\right)^{1/2} 2g\,|\alpha_{N,max}|\left(\frac{8\pi}{3}\right)^{1/2}[u_k e^{-i\omega_r t} + \text{c.c.}] , \qquad (8.26)$$

where $g = (2\pi/3)^{1/2}(1.38)^{-1} = 1.05$, and $\omega_r \simeq (1-\dot{\phi})\,\omega_N' \simeq \omega_N'/4$. The ponderomotive force acting on an electron can be calculated in the usual way:

$$F_{pm}^{(N)} = -\frac{\partial}{\partial z}V_{pm} \simeq \frac{2\pi e^2}{6 m_e}4g^2\left(\frac{N}{V}\right)|\alpha_{N,max}|^2 \simeq 8\,10^8\,\mathrm{eV/\,cm}. \qquad (8.27)$$

Turning now our attention to the plasmas of the Pd-deuteride (whose loading ratio D/Pd shall be denoted, as is costumery, by x), we consider first the electrons' plasma, that we shall simply describe as a plasma of delocalized s-electrons.

The "realistic" plasma frequency is $\left[\left(\frac{N}{V}\right)\ \text{is, as usual, the Pd-density}\right]$

$$\omega'_{se} = \frac{e}{(m_e)^{1/2}} \left(\frac{xN}{V}\right)^{1/2} (1.38) \simeq x^{1/2} 15.2\,\text{eV}, \qquad (8.28a)$$

and the size of the associated CD is

$$\lambda_{se} = \frac{2\pi}{\omega_{se}} = x^{-1/2} 830\,\text{Å}. \qquad (8.28b)$$

The ponderomotive force on the surface electron is

$$F_{pm}^{(se)} = \frac{e^2}{2m_e} \times \left(\frac{N}{V}\right) (0.8)^2 \frac{2\pi}{3} = x\ 2.5 \cdot 10^7\ \text{eV/cm}, \qquad (8.29)$$

quite smaller than both (8.22) and (8.27).

The energetics of this plasma comprises three contributions that arise from the filling of the electron gas states up to the maximum momentum

$$p_F = \left(3\pi^2\ \frac{N}{V}\right)^{1/3} x^{1/3}, \qquad (8.30)$$

the electrons' Coulomb repulsion and the energy gain associated with the coherent plasma oscillations (see Section (5.3)).

The first contribution E_{Pauli} is simply given by (N is the number of Pd atoms)

$$\frac{E_{Pauli}}{N} = \left(\frac{V}{N}\right)^2 \int^{p_F} \frac{d^3\vec{p}}{(2\pi)^3}\ \frac{\vec{p}^2}{2m_e} = \epsilon_{Pauli}\ x^{5/3}, \qquad (8.31a)$$

where

$$\epsilon_{Pauli} = \frac{3}{10}\ (3\pi^2)^{2/3} \left(\frac{N}{V}\right)^{2/3} \frac{1}{m_e} = 3.52\,\text{eV}. \qquad (8.31b)$$

As for the Coulomb contribution, this can also be computed fairly easily:

$$\frac{E_C}{N} = \frac{1}{2}\ x^2 \left(\frac{N}{V}\right) \int d^3x\ V_C(\vec{x}), \qquad (8.32)$$

where

$$V_C(\vec{x}) = \frac{\alpha}{|\vec{x}|}\ e^{-\frac{|\vec{x}|}{\lambda_D}}, \qquad (8.33)$$

is the Debye-screened Coulomb potential, and λ_D is the Debye-screening length. Thus we have

$$\frac{E_C}{N} = \epsilon_C \, x^2, \tag{8.34a}$$

with

$$\epsilon_C = \frac{e^2}{2} \left(\frac{N}{V} \right) \lambda_D^2 \simeq 9.2 \, \text{eV}, \tag{8.34b}$$

if we take $\lambda_D \sim \dfrac{a}{2} = \dfrac{1}{2} \left(\dfrac{V}{N} \right)^{1/3} = 1.27$ Å, a very reasonable value.

As for the energy gain from the coherent plasma, from the results of Section (5.2) we have

$$\frac{E_{se}}{N} = -\epsilon_{se} \, x^{3/2}, \tag{8.35a}$$

with

$$\epsilon_{se} = 2.3 \, \text{eV}. \tag{8.35b}$$

The total energy

$$\frac{E_e}{N} = \frac{E_{Pauli} + E_C + E_{se}}{N} \tag{8.36}$$

as a function of x is plotted in Fig. 8.1.

Finally, let us analyse the deuteron plasma. The first observation that one can make by looking at the (1.1.0)-plane of the Pd lattice (see Fig. 8.2) is that there are two types of equilibrium positions for the plasma oscillations of the deuterons (or protons): the octahedral and the "tetrahedral" sites.

The physical difference between these two sets of sites can be understood by observing that the plasma oscillations of the d-electrons affect the tetrahedral sites much more than the octahedral sites. It is easy to show that the d-electrons' oscillations, of maximum amplitude $|\vec{\xi}|_{max} \simeq 0.6$ Å (see (8.20)) and oriented between neighbouring atoms (in the ξ-direction of Fig. 8.3), concentrate in disks oriented in the orthogonal direction (η) a **static** charge that produces the periodic potential drawn in Fig. 8.4.

Which configuration will the deuterons arrange themselves into depends solely on the energetics of the two different plasmas, that we shall call β-plasma (octahedral sites) and γ-plasma (tetrahedral sites). In both cases we shall be dealing with a new type of plasma configuration — an "excited plasma" — for which the value of the conserved quantity Q (see Eq. (5.29a)) is not negligible but is equal to $\alpha_0^2 \simeq \alpha_{max}^2$.

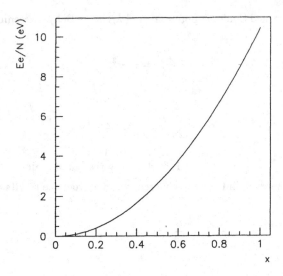

Fig. 8.1. The energy of the s-electron plasma as a function of the loading ratio

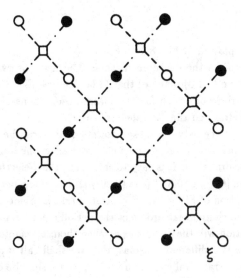

Fig. 8.2. The (1.1.0)-plane of the free lattice of Pd. The empty circles denote the Pd-nuclei, the full circles the octahedral sites, and the empty squares the "tetrahedral" sites.

From Eqs. (5.31) and (5.41) it is easy to see that in this case we have two different solutions according to whether $g \gtrless g_c = 0.77$. For $g < g_c$, calling $y = \frac{\alpha_{max}}{A}$, we have

$$g = 2(y - y^3), \tag{8.37}$$

with the corresponding energy gain with respect to the PGS

$$\frac{\Delta E}{N_D \omega_D} = \frac{2\alpha_{max}^2}{3y^2 - 1} \left[1 - 4y^2 + 3y^4\right]. \tag{8.38}$$

For $g > g_c$, on the other hand, one obtains

$$y \simeq \frac{1}{2g}, \tag{8.39}$$

and

$$\frac{\Delta E}{N_D \omega_D} = -2g^2 \, \alpha_{max}^2. \tag{8.40}$$

Let us now apply this analysis to the β-plasma, whose ideal plasma frequency is

$$\omega_{Dp} = \frac{e}{(m_D)^{1/2}} \left(\frac{N}{V}\right)^{1/2} x^{1/2} = \omega_0 \, x^{1/2} \tag{8.41a}$$

$$\omega_0 = 0.16 \, \text{eV}, \tag{8.41b}$$

which determines the CD size as

$$\lambda_{Dp} = x^{-1/2} \frac{2\pi}{\omega_0} = x^{-1/2} \, 7.85 \cdot 10^{-4} \, \text{cm}. \tag{8.42}$$

We must now evaluate the real plasma frequency ω_β. In order to achieve this we argue as follows: the negative charge density the deuterons find themselves into while in the octahedral sites can be thought of as being constituted by the neutralizing electrons of density $x \left(\frac{N}{V}\right)$ and by the tails of the oscillating d-electrons of approximate density $\delta \left(\frac{N}{V}\right)$. Thus we may write

$$\omega_\beta \simeq \omega_0 \, (\delta + x)^{1/2} \, 1.38 \,, \tag{8.43a}$$

where we have corrected the real plasma frequency by the usual factor 1.38. The associated CD size is:

$$\lambda_\beta = \frac{2\pi}{\omega_\beta} = \frac{2\pi}{1.38\omega_0}(\delta + x)^{-1/2} = 5.7 \cdot 10^{-4} \, (\delta + x)^{-1/2} \, \text{cm} \,. \tag{8.43b}$$

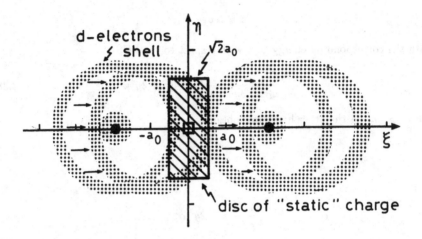

Fig. 8.3. Possible plasma oscillations of the d-shells.

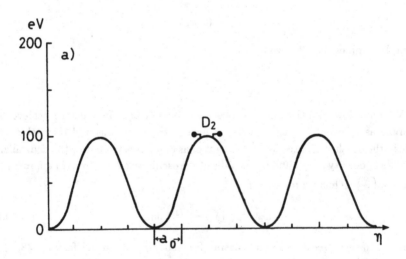

Fig. 8.4. The profile of the electrostatic potential in the η-direction.

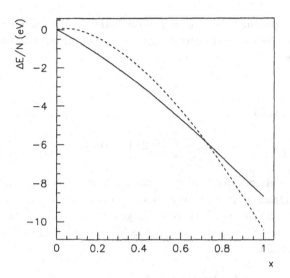

Fig. 8.5. The energy gains $\frac{\Delta E}{N}$ in the β-phase (full line) and in the γ-phase (dashed line)

Eq. (8.43a) allows us to compute the coupling constant

$$g = \left(\frac{2\pi}{3}\right)^{1/2} \left(\frac{\omega_{Dp}}{\omega_\beta}\right) = \frac{1}{1.38} \left(\frac{2\pi}{3}\right)^{1/2} \left(\frac{x}{\delta + x}\right)^{1/2} . \tag{8.44}$$

The coherent dynamics of the β-plasma is completely determined once we estimate the maximum oscillator amplitude :

$$\alpha_{max}^2 = (m_D \omega_\beta)\eta_\beta^2 , \tag{8.45}$$

where $\eta_\beta \simeq 0.4 \overset{\circ}{A}$ appears as a reasonable value. Using now (8.38) and (8.39) together with (8.44), and setting $\delta \simeq 1$ (another reasonable value), we readily obtain the energy gain $\frac{\Delta E_\beta}{N}$ reported in Fig. 8.5.

It is interesting to calculate with the same set of parameters the ponderomotive force that the β-plasma exerts on the electrons at the surface of Pd. Using the well known formula we obtain

$$F_{pm}^{(\beta)} = \frac{e^2}{2m_e} \left(\frac{2\pi}{3}\right) \left(\frac{N}{V}\right) x A^2(x) , \tag{8.46}$$

of the same order of magnitude as $F_{pm}^{(de)}$, see Eq.(8.22) and $F_{pm}^{(N)}$, see Eq.(8.27).

As for the γ-plasma, the frequency of oscillation of the deuteron in the "deep-holes" generated by the linear oscillations of the d-electrons' plasma can be easily evaluated to be

$$\omega_\gamma \simeq 0.65 \, \text{eV}, \tag{8.47a}$$

with an associated CD size

$$\lambda_\gamma = \frac{2\pi}{\omega_\gamma} \simeq 2 \cdot 10^{-4} \, \text{cm} . \tag{8.47b}$$

We expect the γ-plasma to considerably deviate from an ideal plasma, in particular its high delocalization — its motion spans all the extent of the charged disks — makes it very likely that its electric dipole is to a large extent screened by the s-electron gas so that the effective plasma charge e_{eff} is rather smaller than e. This allows us to write for the coupling constant g:

$$g = \left(\frac{2\pi}{3} \right)^{\frac{1}{2}} \left(\frac{\omega_0}{\omega_\gamma} \right) \left(\frac{e_{eff}}{e} \right) x^{\frac{1}{2}} = 0.356 \left(\frac{e_{eff}}{e} \right) x^{\frac{1}{2}} , \tag{8.48}$$

which shows that the γ-plasma is a weakly coupled "excited plasma". In the limit of small g (8.37) and (8.38) yield for the energy gain:

$$\frac{\Delta E_\gamma}{N} = -x\omega_\gamma(m\omega_r)\eta_\gamma^2 \, g = -\epsilon_\gamma x^{\frac{3}{2}} , \tag{8.49a}$$

with

$$\epsilon_\gamma = \omega_\gamma(m\omega_r)\eta_\gamma^2 0.356 \left(\frac{e_{eff}}{e} \right) \simeq 15 \, \text{eV}, \tag{8.49b}$$

where we have set $\eta_\gamma \simeq 1 \text{Å}$ and $\frac{e_{eff}}{e} \simeq 0.20$, values that appear both quite reasonable. Eq. (8.49), however, is not the only contribution to the energy of the γ-plasma configuration: the concentration of charge in the disk-like regions, induced by linear oscillations of the d-electrons' plasma, substantially reduces the potential binding the deuterons to the octahedral sites. As a result the energy of the PGS from which there arises the new coherent process that fills the tetrahedral sites is decreased by the amount

$$\frac{\Delta E_t}{N} = -x\frac{1}{2}m_D(1.38\omega_0)^2\eta_\beta^2 = \epsilon_t x \tag{8.50a}$$

$$\epsilon_t \simeq 4.6 \, \text{eV}, \tag{8.50b}$$

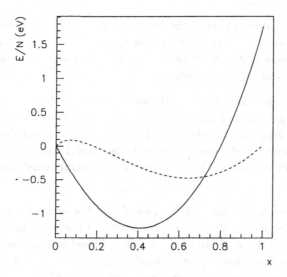

Fig. 8.6. The full energetics of the β- (full line) and γ- (dashed line) phases.

which is nothing else than the height of the harmonic plasma potential at the maximum oscillation amplitude $\eta_\beta \simeq 0.45\text{Å}$ of the β-plasma. The total energy of the γ-plasma is thus expressed by

$$\frac{\Delta E_\gamma}{N} = -\epsilon_\gamma x^{\frac{3}{2}} + \epsilon_t x \ , \tag{8.51}$$

which is compared with the energy of the β-plasma in Fig. 8.5.

As for the ponderomotive force of this plasma acting on a surface electron, it is given by the formula

$$F_{pm}^{(\gamma)} = \frac{e^2}{2m_e} \left(\frac{2\pi}{3}\right) \left(\frac{N}{V}\right) x \frac{m\omega_\gamma}{2}\eta_\gamma^2 = 8.53 \cdot 10^8 \ x \ \text{eV/cm} \ , \tag{8.52}$$

which is to be compared with $F_{pm}^{(\beta)}$ (eq.(8.46)).

Finally, by adding the s-electrons' plasma energy $\frac{E_s}{V}$ to the energy of the deuteron plasma in the β- and γ-phase we obtain the energetics of the two plasmas reported in Fig. 8.6.

While admittedly crude, our analysis has provided us with the conceptual tools to understand a number of subtle phenomena that happen when hydrogen and its isotopes, either molecular or ionized, reach the surface of Pd. Let us summarize them:

(i) the large barriers (several tens of eVs) that must be overcome in order to allow molecular hydrogen to enter the Pd lattice in the ionic state, as experimentally observed, are easily seen to be due (see Section (8.2)) to the "evanescent wave" configurations of the coherent e.m. fields associated with the d-electrons' plasma and the plasma of the Pd ions;

(ii) the double layer that forms at the surface of the cathode can be understood as being the result of the action of ponderomotive forces associated with all plasmas forming in the Pd-deuteride. In particular the contributions of the β- and γ-plasmas to the ponderomotive forces show a strong x-dependence that agrees with the observed fact that the cathode overpotential does increase with loading;

(iii) the different behaviours as a function of x of the energetics of the β- and γ-plasma give us comfortable room for describing the stable phase as a β-phase (deuterons in the octahedral sites) and the γ-phase(deuterons in the tetrahedral sites) as a phase that can be accessed for $x > x_\beta \simeq 0.6$ by working upon the system; and the double layer appears in this respect of crucial importance.

As we can see, the phenomena of CF appear to involve a rather complicated, definitely non trivial set of conditions that the plasmas of the Pd-deuteride seem to "miraculously" fulfill. It is thus not a fortuitous coincidence that the idea of trying CF in this system occurred to people well aware of the reality of a number of such "miracles".

8.4 The three "miracles" of Cold Fusion[i]

At the end of Section (8.1) the hurdles that CF phenomenology puts in the way of any serious theoretical explanation are summarized in the unavoidable task of bringing within the realm of scientific rationality three "miracles", an undertaking that in the GACMP has been shown to be impossible. We shall now see that within the framework of QED coherence the situation is totally different.

Let us begin with miracle #1, namely the incredible suppression by some 40 orders of magnitude of the Coulomb barrier to DD-fusion. As well known, due to the very short-range character of nuclear forces, in order for the DD system to undergo a nuclear fusion process it is necessary that the two deuterons approach at nuclear distances r_N, of the order of a few Fermis. In nuclear physics this is customarily achieved by accelerating a deuteron beam and sending it to a deuteron target, while in stars the kinetic energy necessary to overcome the Coulomb repulsion of the two equal

---

[i]This Section requires the same footnote of Section (8.3)

charges of the deuterons is a consequence of the high temperatures and pressures that exist in their interiors. And in palladium? Nothing of the sort seems to exist, neither accelerators nor high temperatures or pressures, so there is apparently no physical agent that allows the deuterons to surmount the considerable Coulomb barrier that exists at nuclear distances. Thus the almost unanimous verdict of impossibility of CF that came, and still comes from the physics community.

Quantitatively, from the WKB-approximation of the Schrödinger equation, the "overlap" amplitude of the two deuterons at distance r within the classically forbidden region is given simply by the Gamow-amplitude:

$$\eta_G(r) \sim \exp\left\{-(2\mu)^{\frac{1}{2}}\int_r^{r_0} dr'\,[V(r') - E]^{\frac{1}{2}}\right\}\,, \tag{8.53}$$

where $\mu = \frac{m_D}{2}$ is the reduced mass, $V(r)$ is the potential acting upon the two-deuteron system that in the Coulomb case is simply

$$V_c(r) = \frac{\alpha}{r}\,, \tag{8.54}$$

E is the relative energy of the deuterons, and r_0 is the classical turning point, i.e. that value of r for which $V(r_0) = E$.

In the CF case $E \simeq 0$, and if for r_0 one takes $r_0 \simeq 0.7\mathring{A}$, the average distance between the two deuterons in the D_2-molecule, (8.53) can be evaluated explicitly to give

$$\eta_G(r) \simeq \exp -\left\{(2\mu)^{\frac{1}{2}}2\alpha^{\frac{1}{2}}\left[(r_0)^{\frac{1}{2}} - (r)^{\frac{1}{2}}\right]\right\}\,, \tag{8.55}$$

which for $r = r_N \simeq 20F$ implies the infinitesimal value $\eta_G(r_N) \simeq 10^{-60}$. For the fusion rate in the D_2-molecule this implies (taking for the nuclear rate the reasonable value $\Gamma_N \simeq 10^{22}\,\mathrm{sec}^{-1}$) the phenomenally small rate $\Gamma(D_2 \to n\ ^3He + p\ T) \simeq |\eta_G|^2\Gamma_N \simeq 10^{-98}\,\mathrm{sec}^{-1}$. No detectable fusion process can take place with the miserably small η_G given by (8.55)[j].

The only way for fusion to take place in a detectable manner is that for some reason the electrostatic potential seen by the two deuterons is considerably decreased with respect to (8.54): there must be a large, unexpected "screening". From the discussion of the last Section we do not have to look too hard to find that in the γ-phase one of the deuterons belonging to a coherent plasma state sits in a deep potential well, screened by the plasma of d-electrons that has been estimated in Ref. [9] to be

$$eV(r) = -Z_d\frac{\alpha}{2a_0^3}r^2\,, \tag{8.56}$$

---

[j]Actually a more refined calculation increases this result by about 20 orders of magnitude, but this does not change significantly the above conclusions.

with $Z_d = \frac{10}{3}$ and $a_0 \simeq 0.7\text{Å}$, which can be adequately approximated by a constant negative potential $V_0 \simeq -85\,\text{eV}$. The Gamow amplitude for the screened potential $V(r) = \frac{\alpha}{r} + V_0$, $E \simeq 0$ and $r_0 \simeq \frac{\alpha}{V_0} \simeq 1.65\ 10^{-9}\,\text{cm}$, can be readily evaluated:

$$\eta'_G \simeq \exp\left[-(2\mu\alpha r_0)^{\frac{1}{2}}\left(\frac{\pi}{2} - 2\left(\frac{r_N}{r_0}\right)^{\frac{1}{2}}\right)\right] \simeq 10^{-22\pm1} , \qquad (8.57)$$

some twenty orders of magnitude larger than for D_2-tunneling. Naturally, due to the extreme sensitivity of the Gamow amplitude on V_0 and, correspondingly, on r_0 the error in the exponential is at least one order of magnitude. So we see that the rather "stiff" plasma of d-electrons achieves a most remarkable screening of the impossibly high Coulomb barrier (8.54). Miracle #1 appears to be no miracle at all!

As we know from the discussion in Section (8.1), if the enhancement of η'_G [see (8.57)] were all that happened in the Pd lattice, we would obtain DD fusion in the two "incoherent" channels (8.1) and (8.2) with a rate

$$\Gamma_{inc} = |\eta'_G|^2\,\Gamma_N\cdot\left(\frac{N}{V}\right)(x-1) = 6.7(x-1)\ \text{fusions/(cm}^3\ \text{sec)} \qquad (8.58)$$

a very small rate, which is however affected by an uncertainty of at least two orders of magnitude. The factor $(x-1)$ appearing in (8.58) is particularly noteworthy: it stems from the fact that for the screening potential to be effective **it is necessary that all the tetrahedral sites are occupied** ($x = 1$). Thus fusion takes place between the deuterons of the plasma of the γ-phase and the extra deuterons of density $(x-1)(\frac{N}{V})$ that one has been able to accomodate in the lattice and that occupy the "shallow holes" (see Fig. 8.7). Rates of the order of Eq. (8.58) seem to account rather well for experimental observations.

We have seen in Section (8.1), however, that the incoherent rate Γ_{inc}, at which the two usual channels (8.1) and (8.2) are produced, cannot account in any way for the large amounts of excess heat observed by FP: this constitutes what has been termed the miracle #2, that we are now going to discuss. Suppose that in a coherence domain[k] where there are packed xN_{CD} ($N_{CD} \simeq 5.4\ 10^{11}$) deuterons, one has $x > 1$, so that all tetrahedral sites are occupied and CF can take place.[l] Besides the incoherent fusion mechanisms that involve well defined deuteron pairs in the usual *in vacuo* nuclear

---

[k]The coherence domain of the deuteron plasma in the γ-phase, of size $\lambda \simeq 2\cdot10^{-4}\,\text{cm}$ (see (8.47b)), is the basic unit of coherent dynamics. Due to the violent energy releases involved in CF processes, we do not deem it likely that coherence extends well beyond a single CD.

[l]Please note that one of the bonuses of our analysis is that CF does have a threshold value in the loading ratio x. Experimentally $x_{th} < 1$, but this is quite natural, for the measurements refer to the "bulk" x, while the physical threshold refers to the single CD.

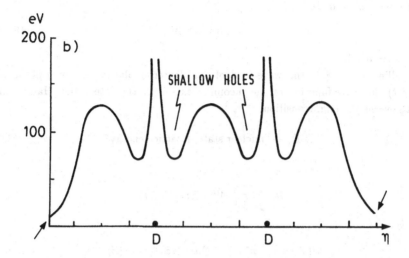

Fig. 8.7. The profile of the electrostatic potential in the η-direction around a deuteron in the γ-phase.

processes, there are important coherent mechanisms that involve **all** the coherent deuterons of the γ-phase.

Let us see whether the collective action of the coherent γ-phase can explain miracle #2. The excess deuteron entering as a plane wave a CD, where the tetrahedral positions are all filled, will have a large number — precisely N_{CD} — of deuterons which he can choose to fuse with. Naturally, in order for coherence to produce the N-enhancements we are well accustomed with, it is absolutely necessary that the large energy releases of fusion processes be **absorbed** by robust, coherent systems that can share large quantities of energy among macroscopic numbers of elementary systems.

Among the coherent e.m. fields that permeate the Pd-lattice there is one which is particularly strong (we have already encountered it in Section (8.2) and Section (8.3)): the e.m. vector potential $\vec{A}_k^{(N)}$ of the plasma of Pd-nuclei, which according to (8.26) is given by

$$\vec{A}_k^{(N)} = 2g\,|\alpha_{max}| \left(\frac{8\pi}{3}\right)^{\frac{1}{2}} \left(\frac{2N}{\omega_N' V}\right)^{\frac{1}{2}} u_k \cos\omega_r t \;, \tag{8.59}$$

where $g \simeq 1.05$, $|\alpha_{max}| \simeq 8.6$ and $\omega_r \simeq \frac{1}{4}\omega_N'$ with $\omega_N' \simeq 0.28\,\text{eV}$. From Eq. (8.59) we can write

$$\vec{A}_k^{(N)} = a_N u_k \cos\omega_r t \;, \tag{8.60a}$$

where we can estimate

$$a_N \simeq 3.2 \ 10^6 \, \text{eV} \ , \tag{8.60b}$$

a very large value.

Calling $\psi_{D_p}(\vec{x}, \vec{\xi}, t)$ the wave-field of the deuteron plasma of the γ-phase and $\eta_D(\vec{x}, t)$ the wave-function of the incoming excess D, the interaction Hamiltonian that governs the e.m. transition

$$\text{D D}_p \rightarrow \text{ nuclear state + lattice energy} \ , \tag{8.61}$$

is

$$H_{int} = e \int_x \vec{A}^{(N)}(\vec{x}, t) \cdot \vec{J}(\vec{x}, t) \ , \tag{8.62a}$$

where

$$\vec{J}(\vec{x}, t) = \int_{\vec{\xi}} \psi_{NS}^*(\vec{x}, \vec{\xi}, t) \vec{J} \psi_{D_p}(\vec{x}, \vec{\xi}, t) \eta_D(\vec{x}, t) \tag{8.62b}$$

is the e.m. current operator,

$$J_k(t) = \langle NS(t)|j_k|\text{D D}_p\rangle \tag{8.62c}$$

is the matrix element for the transition (8.61) and $\psi_{NS}(\vec{x}, \vec{\xi}, t)$ is the wave-field describing the particular nuclear state that gets produced in D D_p fusion. Following our previous discussion one can write:

$$J_k(t) = \langle NS(t)|j_k|NS(t)\rangle \eta_G' \tag{8.62d}$$

expressing the e.m. current as a product of the screened Gamow amplitude and the electromagnetic current generated by the nuclear state evolving (in a region of the dimensions of a few Fermis) under the effect of the coherent electromagnetic field $A_k^{(N)}$ (Eq. (8.60)). Thus $|NS(t)\rangle$ obeys the Schrödinger equation (H_N is the nuclear Hamiltonian)

$$i\frac{\partial}{\partial t}|NS\rangle = (H_N + eA_k^{(N)}j_k)\,|NS\rangle \ . \tag{8.63}$$

We must now solve the preliminary problem of setting up a realistic model for the dynamics of two D's overlapping at nuclear distances, which are from that moment ($t = 0$) on governed by the Hamiltonian $H_N + eA_k^{(N)}j_k$. For the sake of simplicity we set

$$|NS(t)\rangle = c_0(t)e^{-i\omega_0 t}|0\rangle + c_1(t)e^{-i\omega_1 t}|1\rangle + c_2(t)e^{-i\omega_2 t}|2\rangle \tag{8.64}$$

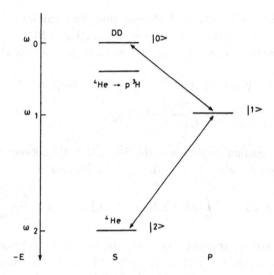

Fig. 8.8. The (idealized) level scheme of the number states (NS) involved in the coherent fusion process.

where the level scheme is depicted in the diagram 8.8.

Calling

$$a = ea_N V_{01} = \langle 1|eA_k^{(N)}j_k|0\rangle \ ,$$ (8.65a)

$$b = ea_N V_{12} = \langle 2|eA_k^{(N)}j_k|1\rangle \ ,$$ (8.65b)

the only nonvanishing matrix elements of the interaction Hamiltonian, it is a simple exercise to show that for times $t \leq \frac{2\pi}{\omega_r}$ (the time that, according to the discussion in Section (3.7), it takes for the nuclear plasma to relax its excess energy), the solutions $c_i(t)$, corresponding to the boundary conditions $c_0(0) = 1$, $c_1(0) = c_2(0) = 0$, are [Ex. ⟨8.2⟩]

$$c_0(t) \simeq 1$$

$$c_1(t) \simeq \frac{a}{\omega_0 - \omega_1}\left(e^{-i(\omega_0-\omega_1)t} - 1\right) \ ,$$ (8.66)

$$c_2(t) \simeq \frac{-ab}{2(\omega_0 - \omega_1)(\omega_1 - \omega_2)}\left(e^{-i(\omega_0-\omega_1)t} - 1\right)\left(e^{-i(\omega_1-\omega_2)t} - 1\right) \ .$$

Considering now that $|\omega_0 - \omega_1| \simeq |\omega_1 - \omega_2| \simeq 10\,\text{MeV}$, we clearly see that $|c_1| \simeq$

$\sqrt{2}|V_{01}|10^{-1}$, and $|c_2| \simeq |V_{01}||V_{12}|10^{-2}$, showing that the dynamics is dominated by the transitions $0 \leftrightarrow 1$, due to the expected sizes of $|V_{01}|, |V_{12}| \simeq 10^{-2}$.

As a result we may express the time independent part of (8.62d) as

$$
\begin{aligned}
eA_k^{(N)}J_k &\simeq ea_N \left[\langle 0|j_k u_k|1\rangle c_1(t)e^{-i(\omega_1-\omega_0)t} + \langle 1|j_k u_k|0\rangle c_1^*(t)e^{-i(\omega_0-\omega_1)t} \right] \eta_G' \\
&= \frac{2a^2}{(\omega_0-\omega_1)}\eta_G' = \frac{2e^2 a_N^2}{(\omega_0-\omega_1)}|V_{01}|^2\eta_G' \ .
\end{aligned}
\tag{8.67}
$$

In the CD the transition amplitude at the time t $(t \leq \frac{2\pi}{\omega_r})$ between the state $|D\,D_p\rangle$ and $|NS\rangle$ can be calculated by using the perturbative formula:

$$
A(D\,D_p \to NS;T) = -i\int_0^T dt \langle NS|H_{int}(t)|D\,D_p\rangle \simeq -iT\eta_G'\frac{2e^2 a_N^2}{(\omega_0-\omega_1)}|V_{01}|^2 N_{CD} \ ,
\tag{8.68}
$$

from which the average transition rate over the time $T \simeq \frac{2\pi}{\omega_r}$ (which, according to the discussion in Chapter 3, represents the time in which the energy stored in the coherent state of Pd nuclei is relaxed to the field fluctuations)

$$
R(D\,D_P \to NS;T) \simeq \frac{|A(D\,D_P \to NS;T)|^2}{T} = -|\eta_G'|^2 \frac{4e^4 a_N^4}{(\omega_0-\omega_1)^2}|V_{01}|^4 \left(\frac{2\pi}{\omega_r}\right) N_{CD}^2 \ .
\tag{8.69}
$$

In order to evaluate the average energy that the fusion process "dumps" onto the lattice, we first notice that under the action of the electromagnetic field $A_k^{(N)}$ the nuclear state $|NS\rangle$ will very rapidly go to the ground state consisting of a coherent state of ^{4}He which will progressively decay into α-particles and migrate **out** of the lattice. There are however other possibilities which may become important in regions where the e.m. field amplitude a_N is smaller (for instance at the boundary of CD's or in the plasma of dislocations, vortices etc.).

Indeed the transition route (see Fig. 8.8) DD $\to$ P-state $\to$ ^{4}He is in general more complicated including the excited ^{4}He-state lying 3.8 MeV below the D D-state.[m] Where a_N is weakened, the dominant e.m. energy producing process, that leads to the ground state ^{4}He, may find some competition from a pure nuclear process such as

$$
^4\text{He}^* \to \text{p T} \ .
\tag{8.70}
$$

We have now all the elements to estimate the power W_{CD} that in each coherence domain is produced by the fusion of the excess $(x-1)N_{CD}$ deuterons with the deuterons

---

[m]As emphasized by Schwinger[10] this state will be able to decay into the p T, but not into the n^{3}He, channel.

of the γ-phase. We obviously have:

$$W_{CD} = (x - 1)N_{CD}R(\text{D } D_P \to NS)\left|\omega_{\text{D D}} - \omega_{4\text{He}}\right| \simeq 10\left|V_{01}\right|^4(x - 1)\text{ Watt/CD} ,$$
$$(8.71)$$

where we have set $|\omega_0 - \omega_1| \simeq 10\,\text{MeV}$. By the further estimates $|V_{01}| \simeq 10^{-2}$ and $(x - 1) \simeq 10^{-1}$ we get the very **rough** evaluation

$$W_{CD} \simeq 10^{-8}\text{ Watt/CD} \tag{8.72a}$$

and, considering that in one cm^3 there are $\frac{1}{(2\,10^{-4})^3}$ CD's, we obtain:

$$W \simeq 1.25\text{ kW/cm}^3 , \tag{8.72b}$$

a very large power, of the magnitude observed by Fleischmann and Pons (see Section (8.1)).

Due to the large uncertainties in the quantities η'_G, and $|V_{01}|$ that affect our calculation of W_{CD}, the above numbers may only be considered as orientative. But even with these *caveats* we clearly see emerging a simple, comfortable explanation of both miracles #2 and #3. Indeed, miracle #2 is simply explained by the numbers appearing in (8.71), while miracle #3, the excess of tritium over neutrons, appears to be a feature of the e.m. coherent path of D D_p-fusion.

Keeping with the spirit of this book, I leave here the fascinating adventure of CF. Even though the discussion has been perforce on a rather crude, semiquantitative level, we seem now to possess the conceptual tools to be able to participate in a meaningful, useful way in this scientific adventure.

Exercises of Chapter 8

$\langle 8.1 \rangle$: Solve the Schrödinger equation (8.11) with the boundary condition $|\chi(0)\rangle = |0\rangle$.

$\langle 8.2 \rangle$: Solve the Schrödinger equation (8.63) in the model space (8.64) with the boundary condition $c_0(0) = 1$, $c_1(0) = c_2(0) = 0$.

References to Chapter 8

1. M. Fleischmann and S. Pons, *J. Electroanal. Chem.*, **261** (1989) 301.
2. J.O'M. Bockris *et al.*, *Fusion Technol.*, **18** (1990) 11.

3. P.K. Iyengar *et al.*, *Fusion Technol.*, **18** (1990) 32.

4. in *The Science of Cold Fusion*, Eds. T. Bressani, E. Del Giudice and G. Preparata, (Società Italiana di Fisica, Bologna, 1991).

5. M. Fleischmann and S. Pons, *Nuovo Cimento* **105A** (1992) 763.

6. E. Botta, T. Bressani *et al.*, *Nuovo Cimento* **105A** (1992) 1663.

7. M. Fleischmann, S. Pons, G. Preparata, *Nuovo Cimento* **107A** (1994) 143.

8. The ideas developed in this and the following Sections appear at a more embryonal stage in T. Bressani, E. Del Giudice and G. Preparata, *Nuovo Cimento* **101A** (1989) 845.

9. G. Preparata, in *The First Annual Conference on Cold Fusion*, (National Cold Fusion Institute, Salt Lake City, 1990).

10. J. Schwinger, in *The First Annual Conference on Cold Fusion*, (National Cold Fusion Institute, Salt Lake City, 1990).

11. G. Preparata, in the Proceedings of ICCF4, Maui (USA), Dec 1993, to appear in *Fusion Technol.*

Chapter 9

QED COHERENCE IN FERROMAGNETISM[a]

9.1   The state of the art in Ferromagnetism

Ever since the dawn of civilization the strange properties of that stone that in classical times could be easily found in the region of Magnesia (Asia Minor) — in modern terms a ferromagnetic piece of matter — have never ceased to appear to man both puzzling and "awesome". Much mystical thought sprang from the observations of the ways magnetic forces act among magnets, and one of the first texts of modern science, Gilbert's "De Magnete"[2], is concerned with the properties of magnets and is still impregnated with considerable mystical and alchemical thought. In modern times magnets have ceased being paradigms of the neoplatonic, animistic conception of the physical world (which however did not play an insignificant rôle in the rise of modern science), and are commonly thought to be well understood within the conceptual framework of the Generally Accepted Condensed Matter Physics (GACMP).

That this commonly held view is far from the truth — and that ferromagnetic behaviour still poses more than one puzzle to the thoughtful and concerned modern scientist — can be best appreciated by quoting some relevant statements made by leading scientists. R.P. Feynman in the final Section of a masterful lecture on magnetism[3] states: "... nor have we explained **why is a loadstone magnetized?** You may say, 'Oh, we just didn't get the right sign'. No, it is worse than that. Even if we **did** get the right sign ...". And in a widely known and appreciated textbook[4], D.L. Goodstein writes: "... Attempts to calculate the effective interaction between magnetic moments in, say, iron tend not even to give the right sign, much less the right magnitude. We are thus in the odd position of being able to explain the properties of a variety of exotic materials but unable to explain why a child's magnet attracts a scrap of iron".

---

[a]This chapter follows closely the paper in Ref. [1].

From these quotations one perceives very clearly that the crucial, fundamental problem of Ferromagnetism is by what QED based mechanism the spins of adjacent electrons of a few transition metals like Fe, Co and Ni, instead of pairing up, as dictated by the Pauli principle (thus neutralizing their magnetic moments) neatly dispose themselves in aligned configurations over **macroscopic** domains (the Weiss domains) that produce the fascinating phenomena of magnetization. A possible such mechanism, that turns out to be classically impossible, was suggested in the early days of quantum mechanics by W. Heisenberg[5] who speculated that the quantum mechanical electrostatic interaction energy of neighbouring electrons might produce an "effective" Hamiltonian H_{eff} (i, j runs over the electrons)

$$H_{eff} = -\sum_{i<j} J_{ij}(\vec{s}_i \cdot \vec{s}_j) \, , \tag{9.1}$$

which depending on the sign of J, the famous "exchange integral", could lead to either ferromagnetic (J-positive) or antiferromagnetic (J-negative) ordering. Since its proposal in the rather remote year 1927, the Heisenberg Hamiltonian (9.1) has been accepted as the correct quantum mechanical basis of ferromagnetism, thus giving a strong impulse to a huge amount of work mainly focused on its statistical mechanical aspects; and in this context one should mention the leading rôle played by its approximation which goes by the name of Ising model. When, however, from the sophisticated phenomenology that has been based on the general properties of the Heisenberg Hamiltonian one turns to the fundamental theoretical problem of computing the sign and the magnitude of J *ab initio* (*i.e.* from the generally accepted picture of the **electrostatic** interactions of condensed matter) the two quotations above bear witness to the fact that our understanding of magnetism is, to say the least, rather precarious. In fact, those who tried to the best of their knowledge of band structure, atomic wave-functions etc. to evaluate J have found that the *naïve* expectations from the Pauli principle are (usually) realized (J is negative) and that the magnitude of J is much too small to secure significant magnetic ordering phenomena.

To summarize, the state of the art of Ferromagnetism is characterized by an extremely sophisticated and successful phenomenology (and technology) on one hand, but by a very unsatisfactory conceptualization on the other, due to the inability of GACMP to find within its electrostatic paradigm simple, natural and fundamental reasons for magnetic order. In the rest of this Chapter we shall see that the theoretical situation becomes much more promising when QED coherence is brought to bear upon the basic mechanisms that correlate electronic spins in a ferromagnetically ordered collective state.

9.2   Coherent electrons' plasmas and the "molecular" field.

As explained in Chapter 5, when embedded in a condensed matter system such as a metal, the electronic "clouds" of the original atoms lose their well known and understood (in elementary quantum mechanics) character to become part of collective quantum mechanical systems, that comprise a **macroscopic** number of elementary objects. These are the Quantum Plasmas (QP) whose QED coherence properties have been studied in Chapter 5.

Let us suppose now that the d-electrons of transition metals, like Fe, Co and Ni, form a coherent plasma, *i.e.* that they can be (approximately) described by a plasma whose coupling constant g is above the critical value 0.77. Let us furthermore assume that their dynamics is well described by the model of Section (5.2). The plasma frequency is thus

$$\omega_{ed} = (1.38) \cdot \frac{e}{(m_e)^{\frac{1}{2}}} \left(\frac{Nd}{V}\right)^{\frac{1}{2}}, \tag{9.2}$$

where the factor 1.38 stems from the expected (atomic) inhomogeneity of the neutralizing ionic charge, and d is the number of d-electrons. Setting, as usual, $\left(\frac{N}{V}\right) = 6.4 \cdot 10^{22}$ cm^{-3} and inserting $d=7$ (Fe) in (9.2) one obtains $\omega_{ed} \simeq 35$ eV, to which there corresponds a coherence domain size

$$\lambda_{ed} = \frac{2\pi}{\omega_{ed}} \simeq 360 \text{\AA}. \tag{9.3}$$

We have seen in Chapter 5 that the plasma oscillations are characterized by a complex unit vector u_k, which determines the character of such oscillations. In particular, for real u_k the oscillations are linear in the direction $\vec{u}$, and they correspond generally to stable configurations, for the complexity of $\vec{u}$ engenders a non-zero angular momentum l for each electron of the plasma ($l^2 \sim [1 - (u_k^* u_k^*)(u_j u_j)]$), that would lead to a local magnetic field of intensity

$$B_C = \frac{e}{2m_e} ld \left(\frac{N}{V}\right)_0 \tag{9.4}$$

($\left(\frac{N}{V}\right)_0$ equals the inverse average volume of the d-shell $\bar{v}_d = \frac{4}{3}\pi (a_0 \mu)^3$, where a_0 is the mean square radius of the shell and μ can be estimated to be $\mu \simeq 1.033$.) The presence of a coherent magnetic field B_C, given by (9.4), would clearly raise the energy density of the system by the quantity $\frac{B_C^2}{2}$, thus making such configuration energetically unfavourable.

However the incompleteness of the d-shell, characteristic of transition metals, introduces a peculiar new dynamical aspect, for the emergence of a coherent magnetic field B_C can be immediately counterbalanced by a local atomic magnetic field in the opposite direction with intensity

$$B_{at} = g_a \frac{e}{2m_e} J \left(\frac{N}{V}\right)_0 = B_C, \tag{9.5}$$

where g_a is the atomic gyromagnetic factor and J is the total angular momentum. The equality of the two magnetic fields (9.4) and (9.5) fixes the size of the angular momentum l as

$$dl = g_a J. \tag{9.6}$$

Note, however, that the value of l is bounded by

$$l_{max} = (m_e \omega_{ed}) \vec{\xi}_{max}^2, \tag{9.7}$$

which for a typical (see Chapter 5) maximum oscillation amplitude $|\vec{\xi}_{max}| \simeq 0.5 \text{Å}$ takes the value $l_{max} \simeq 1$. Thus this mechanism may work only if it so happens that

$$\frac{g_a J}{d} < l_{max} \simeq 1, \tag{9.8}$$

that may well constitute an important criterion for ferromagnetism.

Let us assume that (9.8) is satisfied and that, therefore, the system develops a coherent magnetic field B_C which is completely neutralized by B_{at}. If this were all there happens, we would still have a magnetically neutral system, for the local magnetic moment is just $\frac{e}{2m_e}(dl - g_a J) = 0$. However the establishment of a magnetic B_C, that in view of its compensation by B_{at} is energetically neutral, produces a Zeeman splitting of the electronic levels, that is described by the magnetic Hamiltonian (we take B_C along the z-axis)

$$H_{mag} = (\mu_a)_z B_C = \omega_L J_z, \tag{9.9}$$

where the Larmor frequency is given by

$$\omega_L = \left(\frac{e}{2m_e}\right)^2 \left(\frac{N}{V}\right)_0 g_a^2 J. \tag{9.10}$$

We have thus identified a discrete frequency belonging to the system which depends solely on atomic parameters $\left[\left(\frac{N}{V}\right)_0, g_a, J\right]$ and may possibly lead to a coherent evolution that, as we know well, is always energetically advantageous. Such energetic advantage would then justify (and make it intelligible) the "spontaneous" generation of the "molecular field" B_C.

9.3   Collective evolution of atomic spins in Fe, Co and Ni.

The problem of the QED coherent evolution of atomic spins in the "molecular field" B_C is solved in the by now usual way. We write down the Hamiltonian of the system

$$H = H_{matt} + H_{em}, \qquad (9.11a)$$

where

$$H_{em} = \frac{1}{2}\int_V (\vec{E}^2 + \vec{B}^2)d^3x, \qquad (9.11b)$$

and

$$H_{matt} = \sum_{mm'}\int_V d^3x\,\psi_m^+(\vec{x},t)(H_{at})_{mm'}\psi_{m'}(\vec{x},t)$$

$$+ ig_a\frac{e}{2m_e}\sum_{\vec{k}r}\sqrt{\frac{1}{2V\omega_{\vec{k}}}}\epsilon_{irs}\frac{\vec{k}_r}{|\vec{k}_r|}\epsilon_s^{(r)}(\vec{k})a_{\vec{k}r}e^{-i\omega_{\vec{k}}t}\int d^3x\,e^{i\vec{k}\vec{x}}\psi_m^+(J_i)_{mm'}\psi_{m'} + \text{h.c.}$$

$$\qquad (9.11c)$$

is the wave-field theoretical description of the matter Hamiltonian coupled magnetically to the radiative e.m. field, that is expressed, as usual, in terms of its plane-wave modes' amplitudes $a_{\vec{k}r}$. The wave-fields $\psi_m(\vec{x},t)$ $(-J \le m \le J)$ are labelled by the "magnetic" quantum numbers m, and

$$(H_{at})_{mm'} = \delta_{mm'}\left(\frac{-\vec{\nabla}^2}{2m_e^*} + \omega_L(J + m)\right) \qquad (9.11d)$$

where m_e^* may be related to the "effective" electron mass in the d-shell and where we have subtracted the term $-J\int d^3x\sum_{m=-J}^J \psi_m^+(\vec{x},t)\psi_m(\vec{x},t)$ to normalize to zero the magnetic energy of the lowest lying atomic state $(m = -J)$.

Following the developments of Chapters 2 and 3 we can write down in a straightforward manner the Coherence Equations (CE's) that follow from the Hamiltonian (9.11). However, in order to simplify our problem further it is useful to introduce the average angular momentum operators (as is now customary, we set $\tau = \omega_L t$)

$$J_k(\tau) = \langle \psi(\tau) | \, J_k \, | \psi(\tau) \rangle \, , \tag{9.12}$$

where the state vector $| \psi(\tau) \rangle$ is expressed as

$$| \psi(\tau) \rangle = \sum_m b_m(\tau) \, | m \rangle \, , \tag{9.13}$$

the amplitudes $b_m(\tau)$ being simply related to the amplitudes of the classical wave fields inside the coherence domain associated with the Larmor frequency (of size $\lambda_L = \frac{2\pi}{\omega_L}$) as

$$b_m(\tau) = \frac{1}{\sqrt{N}} \psi_m(\vec{x}, t)|_{\vec{x} \in \text{CD}} . \tag{9.14}$$

It is now quite easy to show that for the angular momenta $J_k(\tau)$ [see Eq. (9.12)] the CE's reduce to the well known Bloch equations, with the fundamental difference, discussed at length in Chapter 2, that for the e.m. field amplitudes we do not make the (generally invalid) slowly varying envelope equation. Thus we may write ($J_\pm = J_1 \pm i J_2$)

$$\dot{J}_3 = -g/2[J_- a^* + J_+ a], \tag{9.15a}$$

$$\dot{J}_+ = g J_3 a^*, \tag{9.15b}$$

$$\dot{J}_- = g J_3 a, \tag{9.15c}$$

$$\dot{a} + \frac{i}{2}\ddot{a} = g J_- , \tag{9.15d}$$

where the coupling constant g is equal to:

$$g = \left(\frac{4\pi}{3} \right)^{\frac{1}{2}} \left[\frac{\frac{N}{V}}{J \left(\frac{N}{V} \right)_0} \right]^{\frac{1}{2}} . \tag{9.16}$$

Following the standard procedure (see Chapter 3) we look for stationary solutions of the system (9.15), whose two constants of motion

$$j^2 = J_+ J_- + J_3^2, \tag{9.17}$$

and

$$Q = |a|^2 + \frac{i}{2}[a^*\dot{a} - \dot{a}^*a] + 2J_3, \tag{9.18}$$

are respectively equal to J^2 and $-2J$. Note that these latter values are appropriate to the "initial" configuration, where the coherent e.m. field is zero and the value of J_3 corresponds to the stable incoherent configuration $-J$.

The linearized short-time equation is the "canonical" $(\tilde{g}^2 = Jg^2)$

$$\frac{i}{2}\dddot{a} + \ddot{a} + \tilde{g}^2a = 0, \tag{9.19}$$

which has runaway solutions, hence coherent behaviour, for

$$\tilde{g}^2 > g_c^2 = \frac{16}{27}. \tag{9.20}$$

Setting

$$J_3 = J\cos\xi, \tag{9.21a}$$

and

$$J_\pm = J\sin\xi e^{\pm i\theta(\tau)}, \tag{9.21b}$$

$$a = Ae^{i\phi(\tau)}, \tag{9.21c}$$

it is a straightforward exercise to find out the values of ξ and the functions, linear in $\tau, \theta(\tau)$ and $\phi(\tau)$ that obey the system (9.15), subject to the conditions (9.17) and (9.18). One obtains $\left(\tilde{A} = \frac{A}{J^{\frac{1}{2}}}\right)$ the equations:

$$\theta + \phi = -\frac{\pi}{2}, \tag{9.22a}$$

$$\dot{\theta} = \tilde{g}\tilde{A}\cot\xi, \tag{9.22b}$$

$$\dot{\phi}^2 - 2\dot{\phi} + 2\tilde{g}\frac{\sin\xi}{\tilde{A}} = 0, \tag{9.22c}$$

$$\tilde{A}^2(1 - \dot{\phi}) + 2(1 + \cos\xi) = 0. \tag{9.22d}$$

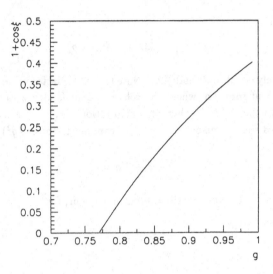

Fig. 9.1. The solution $(1 + \cos\xi)$ of the system (9.22) as a function of the coupling constant $\tilde{g}$

In Fig. 9.1 we plot $(1 + \cos\xi)$, solution of (9.22), as a function of $\tilde{g}$, which from Eq. (9.16) is simply

$$\tilde{g} = \left(\frac{4\pi}{3}\right)^{\frac{1}{2}} \left[\frac{\left(\frac{N}{V}\right)}{\left(\frac{N}{V}\right)_0}\right]^{\frac{1}{2}}. \tag{9.23}$$

The energy gain per atom, ΔE, that one obtains when the system relaxes to the stationary solution (9.22) ($\langle H\rangle_0$ is the energy in the initial configuration where $a = 0$), is

$$\Delta E = \langle H_{matt}\rangle + \langle H_{em}\rangle - \langle H\rangle_0, \tag{9.24}$$

where,

$$\langle H_{matt}\rangle - \langle H\rangle_0 = \frac{1}{2}\left[\frac{B_C}{\left(\frac{N}{V}\right)^{\frac{1}{2}}}\right]^2 (1 + \cos\xi)^2 - i\frac{g}{2}\omega_L(J_+ a - J_- a^*)$$

$$= \Omega\frac{4\pi}{3\tilde{g}^2}\left[\frac{4\pi}{3\tilde{g}^2}(1 + \cos\xi)^2 - 2\tilde{g}\tilde{A}\sin\xi\right], \tag{9.25}$$

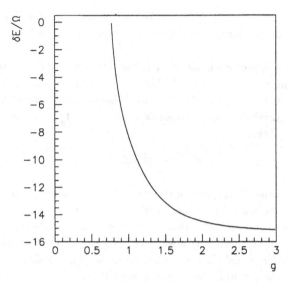

Fig. 9.2. The energy gain $\Delta E/\Omega$ as a function of $\tilde{g}$

and

$$\langle H_{em} \rangle = \frac{\omega_L}{2} A^2 \left[1 - \dot\phi + \frac{\dot\phi^2}{2}\right] = \Omega \frac{4\pi}{3\tilde{g}^2} \tilde{A}^2 \left(1 - \frac{\tilde{g}\sin\xi}{\tilde{A}}\right), \qquad (9.26)$$

where

$$\Omega = \frac{1}{2}(g_a J)^2 \left(\frac{e}{2m_e}\right)^2 \left(\frac{N}{V}\right). \qquad (9.27)$$

In Fig. 9.2 $\frac{\Delta E}{\Omega}$ is plotted as a function of $\tilde{g}$. As we can see, for $\tilde{g} > g_c = \left(\frac{16}{27}\right)^{\frac{1}{2}}$ the energy gain $\frac{\Delta E}{\Omega}$ is negative, showing that the strongly coherent process that develops above the critical coupling does lead to an energetic advantage, thus completely justifying the formation of the magnetic field B_C in the d-electrons' plasma. We should recall at this point that the developments above assume that the constraint (9.8) is satisfied. What happens when it is not? According to our discussion, in such case a macroscopic B-field would appear in the initial configuration (which would be energetically unfavourable) unless the d-electrons' plasma system goes into an **antiferromagnetic** configuration, which can be easily seen to be perfectly possible. Indeed a configuration of alternating angular momenta is totally compatible with the CE's of the stationary plasma, being associated with a coherent e.m. field that instead of being circularly polarized, as it happens in the parallel (ferromagnetic) configuration, is linearly polarized. In this case the theory we have worked out so far goes through essentially unchanged, but for the circumstance that one is now dealing with **antiferromagnetism**, whose discussion, however, is outside the scope of this book.

To get to the important question of the spontaneous magnetization, that gives magnets their peculiar properties, it should be clear that within domains of (approximate) radius $\pi/\omega_L = d_w$ all atoms are polarized in the same direction. From equation (9.9), introducing the appropriate values for $\left(\frac{N}{V}\right)_0 \simeq 4.6 \cdot 10^{23} \mathrm{cm}^{-3}$, $g_a J \sim 5$ and $g_a \simeq 1.5$ we obtain

$$d_w = \frac{\pi}{\omega_L} \simeq 0.23 \text{ mm},\tag{9.28}$$

which determines the (linear) size of the Weiss domains in the range of a fraction of mm, which appears in good agreement with observations.

Furthermore the Bloch walls, that are observed to separate the Weiss domains, according to our theory must have a width corresponding to the size of the coherence domains of the d-electrons' plasma, for the direction of magnetization (the direction of the angular momentum $\vec{l}$) inside a coherence domain of the d-electrons' plasma must remain constant. Thus from (9.3) we derive a typical width of Bloch walls of the order of $400\mathring{A}$, just the observed order of magnitude.

As for the magnetization resulting from the coherent process analysed in the Section (9.1) one simply gets (clearly at $T = 0$, for we have been implicitly working at this temperature)

$$m(0) = (g_a J) \cdot (1 + \cos\xi)\frac{e}{2m_e},\tag{9.29}$$

whose most noteworthy feature is that, differently from the Heisenberg model, the magnetization in units of the Bohr magneton $\left(\frac{e}{2m_e}\right)$ need not be an integer value. By applying (9.29) to Fe, Co, Ni, whose $g_a J$ are 6, 6 and 5 respectively[b], the observed magnetizations require $(1 + \cos\xi)$ to equal 0.369, 0.286 and 0.123 respectively. From Fig. 9.1 these values imply the following values of $\tilde{g}$ at $T = 0$:

$$\tilde{g}(0)_{Fe} = 0.966,$$

$$\tilde{g}(0)_{Co} = 0.906,\tag{9.30}$$

$$\tilde{g}(0)_{Ni} = 0.822,$$

which from (9.16) in turn determine the d-shell root mean-square radii as follows

---

[b]We are obviously assuming that the d-shell electrons follow all three Hund's rules, as it happens in the atom.

$$(a_0)_{Fe} = 0.81\text{Å} \ (0.80\text{Å}),$$

$$(a_0)_{Co} = 0.77\text{Å} \ (0.76\text{Å}),$$

$$(a_0)_{Ni} = 0.71\text{Å} \ (0.71\text{Å}), \tag{9.31}$$

that should be compared with the generally accepted estimates[6] reported in parentheses. The agreement is quite remarkable.

9.4 Magnons and the Curie temperature of Fe, Co and Ni

In order to understand what happens at $T \neq 0$, and in particular to determine the Curie temperature T_C at which the spontaneous magnetization is lost, we must analyse the problem of the spectrum of the quantum fluctuations and of their thermal excitation.

According to the discussion of Section (3.6), we are clearly interested in the "incoherent" fluctuations, for these are the constituents of the incoherent fluid, whose buildup at the expense of the coherent fluid is finally responsible for the phase-transition that leads to the loss of magnetic order. Such phase-transition, that merely consists in the loss of magnetization without any modification of the geometrical structure of the solid, is clearly of the second-order type, taking place when the incoherent fluid has completely invaded the coherent one. Thus in order to study the equilibrium configurations of these two fluids as a function of the temperature T, we must first study the spectrum of the incoherent fluid and then determine the energy that one must spend in order to excite a given incoherent quantum fluctuation out of the coherent, magnetically ordered ground state.

The problem of determining the spectrum of the incoherent fluctuations is very easily solved by considering the "free" term, H_0, of the matter Hamiltonian, H_{matt}, appearing in (9.11):

$$H_0 = \sum_{m=-J}^{J} \int d^3x \psi_m^t(\vec{x}, t) \left(\frac{-\vec{\nabla}^2}{2m_e^*} + \omega_L(m + J) \right) \psi_m(\vec{x}, t). \tag{9.32}$$

The structure of H_0, described by Eq. (9.32), is noteworthy for two reasons: first the mass parameter m_e^*, which should be closely related to the effective electron mass in the d-electrons' bands (determining their widths), and second the magnetic energy of the atomic spins in the coherent magnetic field B_C, that is expressed in terms of the Larmor frequency ω_L (Eq. (9.9)). The diagonalization of (9.32) is immediately carried out in terms of the "spin-wave" decomposition of the wave-fields

$$\psi_m(\vec{x}, t) = \sum_{\vec{q}} \frac{1}{(V)^{\frac{1}{2}}} \, \eta_{m\vec{q}}(t) e^{i\vec{q} \cdot \vec{x}}, \qquad (9.33)$$

with the equal-time commutation relations

$$[\eta_{m\vec{q}}(t), \eta_{m'\vec{q}'}(t)] = \delta_{mm'} \delta_{\vec{q}\vec{q}'}, \qquad (9.34)$$

which yields:

$$H_0 = \sum_{m=-J}^{J} \sum_{\vec{q}} \eta_{\vec{q}m}^{+} \eta_{\vec{q}m} \omega_m(\vec{q}), \qquad (9.35)$$

with

$$\omega_m(\vec{q}) = \frac{\vec{q}^2}{2m_e^*} + \omega_L(m + J). \qquad (9.36)$$

This latter equation, however, does not describe the excitation spectrum of the spin-waves (the magnons) out of the magnetically ordered CGS. In order to get the magnon spectrum we must add to (9.36) the gap Δ, separating the CGS from the perturbative ground state, which is simply equal to $\frac{\Delta E}{N}$ (see Fig. 9.2). Thus the magnon spectrum is given by

$$\epsilon_m(\vec{q}) = \omega_m(\vec{q}) + \Delta, \qquad (9.37)$$

which turns out to have just the form $\left(D = \frac{1}{2m_e^*} \right)$

$$\epsilon_m(\vec{q}) = D\vec{q}^2 + \Delta_m, \qquad (9.38)$$

a parametrization usually employed for the magnon spectrum.

The magnons, possessing a boson character, get thermally excited with a Bose-Einstein distribution whose mean occupation numbers are

$$n_{m\vec{q}} = \frac{1}{e^{\epsilon_m(\vec{q})/T} - 1}. \qquad (9.39)$$

Note that the chemical potential has been set to zero, for there is no constraint on the number of magnons present in the system, provided we are below the Curie temperature. Thus the number of magnons $N_m(T)$ as a function of the temperature T is given by:

$$N_m(T) = \frac{V}{(2\pi)^3} \sum_{m=-J}^{J} \int d^3q \frac{1}{e^{\epsilon_m(\vec{q})/T} - 1}. \tag{9.40}$$

The excitation of magnons depletes the CGS, which comprises $N_C(T)$ elementary systems, whose number is obviously given by

$$N_C(T) = N - N_m(T), \tag{9.41}$$

where N is the total number of such systems. As a result the effective coupling constant $\tilde{g}$ [see Eq. (9.19)] gets renormalized as

$$\tilde{g}(T)^2 = \tilde{g}(0)^2 \left[1 - \frac{N_m(T)}{N} \right]. \tag{9.42}$$

The Curie temperature T_C is then given by

$$\tilde{g}(T_C)^2 = \frac{16}{27}, \tag{9.43}$$

at which value the spin system ceases to be strongly coherent. Performing the integration involved in Eq. (9.40), using the gap $\Delta(\tilde{g})$, plotted in Fig. 9.2, and substituting the observed Curie temperatures T_C for Fe, Co and Ni respectively, we can determine the "stiffness parameters" D (we display in parentheses the available experimental information)

$$\begin{aligned} D_{Fe} &= 400 \text{ meV\AA}^2 \ (344 \pm 20), \\ D_{Co} &= 650 \text{ meV\AA}^2 \ (740 \pm 80), \\ D_{Ni} &= 460 \text{ meV\AA}^2 \ (400 \pm 20). \end{aligned} \tag{9.44}$$

Finally from (9.40) and (9.42) we can determine $\tilde{g}(T)$ in the range $0 < T < T_C$, and we are thus able to predict the magnetization curves $m(T)/m(0)$ for Fe, Co, and Ni that are compared with experimental data in Fig. 9.3.

9.5 Outlook

The theory of Ferromagnetism that, based on QED coherence, has been worked out in the preceding Sections is still at a rather preliminary stage. The extension of the

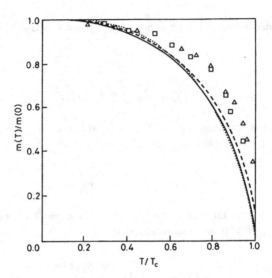

Fig. 9.3. The temperature dependence of the relative magnetization $m(T)/m(0)$ for Fe (dashed line), Co (solid line) and Ni (dotted line) compared with experimental results: $\triangle$ (Fe), $\square$ (Co, Ni).

field of ferromagnetic (and antiferromagnetic) phenomena is so great that much more work is needed before we can appreciate the full potential of the approach sketched in this Chapter. However, what I believe we have definitely achieved is to surmount the obstacle that so far has blocked a fundamental understanding of ferromagnetism, namely a reliable *ab initio* determination of the exchange integral J, that could fit the data, and in particular the large Curie temperatures of Fe, Co and Ni.

Rather than engaging ourselves in the seemingly inane effort to understand J, the possibility has been explored that the main features of Ferromagnetism might arise from some kind of QED coherent behaviour, whose relevance in Condensed Matter Physics is the subject of this book. We have found that a very natural mechanism for generating a ferromagnetic state arises from assuming a QED coherent dynamics for the d-electrons' plasma, that gives rise to a local "molecular" magnetic field B_C, which in turn supplies a discrete Larmor frequency ω_L that triggers a coherent dynamical evolution of the atomic magnetic moments. In this way a non-zero macroscopic magnetization emerges over coherence domains — the Weiss domains — whose size $d_w = \frac{\pi}{\omega} \simeq 0.23$ mm (see Eq. (9.28)) is of the right order of magnitude. But not only the size of the Weiss domains checks with observations, also the thickness of the Bloch walls, according to Eq. (9.3), is of the right order of magnitude.

Getting now to the size of the spontaneous magnetization, unfortunately it depends in a very sensitive way on the root mean square radius a_0 of the electrons'

d-shell, for which only estimates exist[6] (which, however, are deemed rather reliable). As one can see from Eq. (9.31) the agreement with such estimates is indeed quite remarkable, a fact that should not be undervalued. It should also be stressed that our theory handsomely avoids one of the difficulties of the Heisenberg model, namely its necessarily integral (in units of μ_B) magnetic moments.

As for the Curie temperatures, again a precise determination requires a very good knowledge of the spectrum of the spin-waves (the magnons). Here also we have chosen to test our theory by determining the stiffness parameters D from the observed Curie temperatures T_C and, as shown by Eq. (9.44), the values we get are rather reasonable. Finally the magnetization curves of Fig. 9.3, which are now parameter free, in view of the rudimentary character of our analysis appear acceptable.

In leaving this Chapter on Ferromagnetism at this junction I am acutely aware of the immense amount of work, both experimental and theoretical, that has been completely left behind. However, the outlook appears so promising that the fragmentary nature of these developments can perhaps be forgiven.

References to Chapter 9

1. E. Del Giudice, B. Giunta and G. Preparata, *Il Nuovo Cimento* **14D** (1992) 1145.
2. W. Gilbert, *De Magnete* (Dover, New York, 1958).
3. R. P. Feynman, R. B. Leighton and M. Sands, *Lectures on Physics - Vol.2* (Addison-Wesley, London, 1969), p.37-13.
4. D. L. Goodstein, *States of Matter* (Dover, New York, 1985), p.422.
5. W. Heisenberg, *Z. Phys.* **49** (1928) 619.
6. O. K. Andersen and O. Jepsen, *Physica* **B91** (1975) 317.

Chapter 10

DYNAMICS AND THERMODYNAMICS OF WATER[a]

10.1 The many puzzles of liquid water

Liquid water is traditionally one of the outstanding challenges to modern science, owing to the large number of its puzzling anomalies[2], that as a matter of fact turn out to be all essential to the existence of life on our planet. Such anomalies include the behaviours of its density, compressibility and viscosity at low temperature and pressure, suggesting that in such thermodynamical conditions each molecule tends to increase the volume it occupies, like it happens for the molecules of ice. This tendency prompted Röntgen[3] to propose a two-phase model of water, in which chunks of "ice-like" matter are dissolved in a fluid that behaves normally.

After the initial condemnation of Röntgen's views by Bernal and Fowler[4], who pointed out that the quantum mechanical identity of water molecules could not support the flagrant disregard of the Ockham's razor that Röntgen's model implied, in more recent times there have been several attempts to revive this intuitive model within the generally accepted frameworks of dynamics and thermodynamics. Such frameworks allowing only the consideration of electrostatic forces, the main dynamical element in these attempts has been the highly directional force usually called H-bond.

This peculiar electrostatic force, whose fundamental origin is still rather mysterious, is assumed to be described by an angle-dependent two-body potential $V_{HB}(\vec{x})$, whose action is maximum at well defined distances and angles, *i.e.* where the bond is "intact", otherwise its effects become negligible, and the bond is "broken", allowing the molecules to approach at closer distances. Using the tools of statistical mechanics many researchers[5-7] have investigated a dynamical scenario where each pair of molecules has a probability p_{HB} to be connected by an H-bond, whose lifetime is

---

[a]This chapter follows closely the paper in [1]

τ_{HB}. As a result a "flickering" low-density gel-like structure of H-bonded molecules coexists with the unbounded molecules that nest in the cages of the structure and in its interstices. And this, surprising though it may appear, is the picture that is now generally accepted.

Another intriguing aspect of the anomalies of water is the sharp increase of many physical observables (compressibility, specific heat, viscosity) that occurs in super-cooled water at around $-45°C$, where the (extrapolated) density of supercooled water falls to the density of ice $\dfrac{\rho_{ice}}{\rho_{water(4°C)}} = 0.92^8$. This seems to suggest that the "ice-like structure", required to explain the anomalies, should "invade" the liquid just at that temperature.

It must be stressed at this point that the concept of H-bond is merely phenomeno-logical, a useful device to classify and organize the data. In order to simulate the dynamics of water effective molecular pair potentials, alluded to above, have been introduced which, in the framework of the computer Monte Carlo simulations of Molecular Dynamics, could be chosen so as to mimic the phenomenological H-bond array of water[9]. In this way the unsolved dynamical problem of liquid water has been transferred from the H-bond to the origin of those pair potentials. The first question, however, is whether condensed matter can be described in terms of pair potentials only. Stanley and Teixeira (p. 2 of Ref. [6]) for instance point out that "... a description of molecular behaviour of water by an effective pair potential will never be completely realistic, because of the existence of many-body forces and the complexity of water."

Another question is how the water molecule, in its **electronic ground state**, can possibly allow the emergence of force fields that give rise to the highly directional H-bonds. Where are in the electron cloud of H_2O (D_2O) **in the ground state**, the protruding "hooks", the "rabbit's ears" or, technically, the lone pairs required for H-bonds? It is well known, in fact, that the electron cloud of the H_2O ground state has a smooth, potato-like structure, so that in Ref. [10] J.D. Buckingham observes that "... it appears that the hydrogen bond can be understood in terms of an **electrostatic** attraction, implying that distortion of the electronic structure is not important." But, then, how is it possible that an electrostatic interaction emerges in a highly mobile environment?

In conclusion, a theoretical analysis of the generally accepted phenomenological descriptions appears to demand that, at least, the molecules in the condensed phase are in a ground state that differs from the ground state of the isolated molecule. In this new ground state their electrons' clouds would then be so distorted as to produce the observed binding, and to possibly increase the values of molecular parameters such as the effective radius and the dipole moment, as suggested by phenomenological

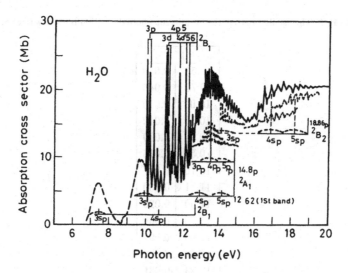

Fig. 10.1. Photoabsorption spectrum of water vapor (see Gurtler, Saile and Koch, Ref. [11])

analyses.

In this Chapter we shall employ the ideas of QED coherence to have a fresh look at the puzzling phenomenology of water. Our novel point of view will make it natural to look at liquid water not as a "molten ice" but rather as a "condensed vapor". Thus we consider the evolution of an assembly of initially independent molecules that by the QED instability of their electrons' ground states are driven toward a coherent dynamical state, where large intermolecular attractions emerge, capable of accounting for the observed thermodynamics of water as well as for the actual dynamical evolution of the vapor-liquid phase transition.

10.2 The coherent ground state (CGS) of water

As well known the symmetry of the water molecule is the point group C_{2v}, which possesses four irreducible representations, that are labelled as A_1, A_2, B_1, B_2. The molecular orbitals having the corresponding symmetry are denoted a_1, a_2, b_1, b_2 respectively. The electronic ground state is, following standard notation, $(1a_1)^2(2a_1)^2(1b_2)^2(3a_1)^2(1b_1)^2$.

The electronic transitions we are interested in are studied, for instance, in Ref. [11], whose experimental results are reported in Fig. 10.1. Information about the oscillator strengths of these transitions between the ground state and the excited states,

Table 10.1. The oscillator strengths of different photoabsorption lines of water vapor

E(eV)	f
7.50	0.050
9.70	0.045
10.00	0.010
10.17	0.012
10.36	0.009
10.54	0.008
10.76	0.007
11.11	0.030
11.50	0.020
11.76	0.030
11.90	0.003
12.06	0.022
12.23	0.031
12.50	0.023

reported in Fig. 10.1, can be found in Ref. [11]. In Table 10.1 we report these data. The estimated accuracy of the numbers of Table 10.1 is about 20 % (10% for the transition at 7.5 eV).

Let us now write the CE's (see Chapter 3) for the transition between the ground state and a generic excited state, within a CD (i.e. neglecting any space-dependence). Setting ($m = 1, 2, 3$)

$$b_0(t) = \beta_0(t) e^{\frac{i\omega_0}{2}t}$$
$$b_1^m(t) = \beta_1^m(t) e^{\frac{-i\omega_0}{2}t}$$

(10.1)

where ω_0 is the frequency of the transition, and introducing the dimensionless time $\tau = \omega_0 t$, the CE's become

$$\dot{\beta}_0(\tau) = -i \sum_m \frac{e\delta}{\omega_0} \frac{1}{(2\omega_0)^{\frac{1}{2}}} \left(\frac{N}{V}\right)^{\frac{1}{2}} \left(\frac{\omega_0}{2m_e}\right)^{\frac{1}{2}} \sum_r \int d\Omega_{\underline{k}} \epsilon_{\underline{k}r}^{*m} a_{\underline{k}r}^* \beta_1^m(\tau) \qquad (10.2a)$$

$$\dot{\beta}_1^m(\tau) = -i \frac{e\delta}{\omega_0(2\omega_0)^{\frac{1}{2}}} \left(\frac{N}{V}\right)^{\frac{1}{2}} \left(\frac{\omega_0}{2m_e}\right)^{\frac{1}{2}} \sum_r \int d\Omega_{\underline{k}} \epsilon_{\underline{k}r}^m a_{\underline{k}r} \beta_0(\tau) \qquad (10.2b)$$

$$\dot{a}_{\vec{k}r} + \frac{i}{2}\ddot{a}_{\vec{k}r} + i\mu a_{\vec{k}r} = -i\frac{e\delta}{\omega_0(2\omega_0)^{\frac{1}{2}}}\left(\frac{N}{V}\right)^{\frac{1}{2}}\left(\frac{\omega_0}{2m_e}\right)^{\frac{1}{2}}\sum_m \beta_1^m \beta_0^* \epsilon_{\vec{k}r}^{m*}. \tag{10.2c}$$

Defining

$$\alpha_m(\tau) = \left(\frac{3}{8\pi}\right)^{\frac{1}{2}}\sum_r \int d\Omega_{\vec{k}} a_{\vec{k}r}\epsilon_{\vec{k}r}^m, \tag{10.3}$$

the CE's (10.2) can be written;

$$\dot{\beta}_0(\tau) = -i\sum_m g\alpha_m(\tau)^*\beta_1^m(\tau), \tag{10.4a}$$

$$\dot{\beta}_1^m(\tau) = -i\, g\alpha_m(\tau)\beta_0(\tau), \tag{10.4b}$$

$$\dot{\alpha}_m(\tau) + \frac{i}{2}\ddot{\alpha}_m(\tau) + i\mu\alpha_m(\tau) = -i\, g\beta_0^*(\tau)\beta_1^m(\tau), \tag{10.4c}$$

where we have introduced the coupling constant

$$g = \left(\frac{2\pi}{3}\right)^{\frac{1}{2}}\frac{\omega_p}{\omega_0}\delta = \left(\frac{2\pi}{3}\right)^{\frac{1}{2}}\left(\frac{\omega_p}{\omega_0}\right)^{\frac{1}{2}}f_{01}^{\frac{1}{2}}, \tag{10.5}$$

where f_{01} is the oscillator strength of the transition $0 \longleftrightarrow 1$, and the plasma frequency is equal to $\omega_p = \frac{e}{(m_e)^{\frac{1}{2}}}\left(\frac{N}{V}\right)^{\frac{1}{2}}$ (m_e is the electron mass). μ, — the "photon mass" — has the following expression

$$\mu = \frac{e^2}{\omega_0}\left(\frac{N}{V}\right)\lambda = -\frac{3}{2}\left(\frac{\omega_p}{\omega_0}\right)^2\sum_n f_{nb}\frac{\omega^2}{(E_n - E_b)^2 - \omega^2}, \tag{10.6}$$

where the oscillator strength f_{bn} for the transition $b \longleftrightarrow n$ is given by

$$f_{bn} = \frac{2}{3}\frac{m_e}{\omega|E_n - E_b|}\sum_j |\langle b|\, J_j\, |n\rangle|^2, \tag{10.7}$$

and ω is the frequency of the e.m. field-mode.

As usual the CE's (10.4) admit the following constants of motion:

$$|\beta_0(\tau)|^2 + \sum_m |\beta_m(\tau)|^2 = 1, \tag{10.8}$$

and

$$\sum_m \{2|\alpha_m|^2 + i[\dot{\alpha}_m\alpha_m^* - \alpha_m\dot{\alpha}_m^*] + |\beta_1^m|^2\} - |\beta_0|^2 = \Delta. \tag{10.9}$$

Eq. (10.8) represents the conservation of the total number of particles, while Eq. (10.9) enforces momentum conservation. As usual, the constant Δ can be obtained by noticing that, in the initial configuration, *i.e.* at $\tau = 0$, neglecting terms $O\left(\frac{1}{N}\right)$, $\alpha_m(0) = \dot{\alpha}_m(0) = 0$, thus

$$\Delta = \sum_m |\beta_1^m(0)|^2 - |\beta_0(0)|^2 \tag{10.10}$$

is the initial difference of populations of the two electronic levels. In the "cold start" one has $\Delta = -1$.

By differentiating (10.4c) with respect to time, and using (10.4a) and (10.4b) together with the initial conditions (10.10) we get for short times τ:

$$\frac{i}{2}\dddot{\alpha}_m + \ddot{\alpha}_m + i\,\mu\,\dot{\alpha}_m + g^2\alpha_m = 0, \tag{10.11}$$

whose general solution is a superposition of the exponentials $e^{ip_k\tau}(k = 1, 2, 3)$, where p_k is a solution of the cubic equation

$$\frac{1}{2}\,p^3 - p^2 - \mu p + g^2 = 0. \tag{10.12}$$

When (See Section (3.2))

$$g^2 > g_c^2 = \frac{8}{27} + \frac{2}{3}\mu + \left(\frac{4}{9} + \frac{2}{3}\mu\right)^{3/2} \tag{10.13}$$

Eq. (10.12) possesses a pair of complex conjugate solutions. As a consequence the e.m. field amplitudes $\alpha_m(\tau)$ "run away" exponentially: the system leaves this state, that we may call the Perturbative Ground State (PGS, for in this state the radiative e.m. field consists of "zero-point" fluctuations only) and migrates to the true ground state, the Coherent Ground State (CGS).

Before finding the CGS and discussing its properties, let us try to identify the excited level of the water molecule that gives rise to the coherent dynamics whose initial stage is governed by Eq. (10.11). We notice that, due to the proportionality of both g^2 and μ to the density $\left(\frac{N}{V}\right)$ (See Eqs.(10.5) and (10.6)), Eq. (10.13) puts a lower bound on $\left(\frac{N}{V}\right)$ that we may call $\left(\frac{N}{V}\right)_{crit}$, the "critical" density.

In Table 10.2 we report the values of g^2 and μ for each of the "low-lying" levels of the water molecule, together with the value of $d_{crit} = \left(\frac{N}{V}\right)_{crit}$, as determined by Eq. (10.13).

It is clear now that as long as the "vapor" density remains below the smallest of the critical densities that, according to Table 10.2, belongs to the level at 12.06 eV,

Table 10.2. Coupling constants g^2, photon masses μ and critical densities d_{crit} for the water vapor's photo-absorption lines of Table 10.1

E (eV)	g^2	μ	d_{crit} (g/cm^3)
7.50	0.26	−1.10	0.371
9.70	0.14	−1.46	0.314
10.00	0.03	−1.12	0.433
10.17	0.03	−1.04	0.464
10.36	0.02	−1.05	0.463
10.54	0.02	−1.08	0.453
10.76	0.02	−1.26	0.393
11.11	0.07	−1.36	0.350
11.50	0.04	−1.52	0.319
11.76	0.06	−1.42	0.337
11.90	0.01	−1.21	0.410
12.06	0.04	−1.57	0.310
12.23	0.06	−1.12	0.425
12.50	0.04	−1.43	0.339

the system of water molecules remains in the PGS, whence it "runs away" as soon as such critical density is reached. When this happens the e.m. "zero-point" fluctuations with frequency $\omega = 12.06$ eV start to build up and the water molecules will begin to oscillate between the ground state and the excited level at 12.06 eV: all the other levels will be from now on totally ignored by the dynamical evolution of the physical system: water molecules plus e.m. field.

As a consequence our vapor undergoes a phase transition toward the liquid state, and it is highly interesting to note that Table 10.2 gives for the critical density a value that is in quite good agreement with the experimentally observed critical density of water. Indeed we compute a critical molar volume $V_c \sim 57$ cm^3, to be compared with the observed value $V_c = 55.61$ cm^3.

In order to determine the CGS we write the fundamental fields as:

$$\beta_0(\tau) = B_0 e^{i\theta_0(\tau)}, \tag{10.14a}$$

$$\beta_1^m(\tau) = B_1 u_m e^{i\theta_1(\tau)}, \tag{10.14b}$$

$$\alpha_m(\tau) = A u_m e^{i\phi(\tau)}, \tag{10.14c}$$

where the complex vector u_m is normalized to 1. Inside a CD, we consider first space independent fields, and for them Eqs.(10.4) become:

$$iB_0\dot{\theta}_0 e^{i\theta_0} = -igAB_1 e^{i(\theta_1 - \phi)}, \tag{10.15a}$$

$$iB_1\dot{\theta}_1 e^{i\theta_1} = -igAB_0 e^{i(\theta_0 + \phi)} \tag{10.15b}$$

$$iAe^{i\phi}\left[\dot{\phi} - \frac{\dot{\phi}^2}{2} + \mu\right] = -igB_0 B_1 e^{i(\theta_1 - \theta_0)}, \tag{10.15c}$$

where, according to Eq. (10.8), we may set:

$$B_0 = \cos\gamma \quad B_1 = \sin\gamma.$$

We notice that in Eq. (10.15c) the value of μ needs not be the same as in the short-times equation (10.11), which was determined from Eq. (10.6) when both the matter and the e.m. field were in their ground state. For it is clear from Eq. (10.6) that μ depends on the state in which the e.m. field plus matter actually find themselves in. As a consequence the system (10.15) is highly non-linear. From our Ansatz (10.14), the frequency of the e.m. field in the CGS is

$$\omega_r = |1 - \dot{\phi}|\omega_0, \tag{10.16}$$

while the water molecules are in a superposition of the ground state and of the excited state at 12.06 eV, whose energy is E_{GS}. From Eq. (10.6), and neglecting the oscillator strength f_{1n}, the new value of μ — μ_r — becomes:

$$\mu_r = -\frac{3}{2}\left(\frac{\omega_p}{\omega_0}\right)^2\left[\sum_{n=1}^{\infty}\frac{\cos^2\gamma f_{on}\omega_r^2}{(E_n - E_{GS})^2 - \omega_r^2} + \frac{\sin^2\gamma\, f_{10}\,\omega_r^2}{(E_0 - E_{GS})^2 - \omega_r^2}\right]. \tag{10.17}$$

Eq. (10.17) defines μ as a function of E_{GS}, γ and $\dot{\phi}$, which are to be determined by the Eqs.(10.15), that we are now going to solve. It is very easy to see that we have a constraint on the phases

$$e^{i\chi} = e^{i(\phi - \theta_1 + \theta_0)} = \delta = \pm 1, \tag{10.18}$$

which when obeyed leads to

$$\dot{\theta}_0 = -g\delta A\tan\,\gamma, \tag{10.19a}$$

$$\dot{\theta}_1 = -g\delta A\cot\,\gamma, \tag{10.19b}$$

$$\dot{\phi} - \frac{1}{2}\dot{\phi}^2 + \mu_r = -\frac{g\delta}{2A}\sin 2\gamma. \tag{10.19c}$$

These equations must be supplemented by the constants of motion:

$$\dot{\chi} = (\dot{\phi} - \dot{\theta}_1 + \dot{\theta}_0) = 0, \tag{10.19d}$$

and

$$2A^2(1 - \dot{\phi}) - \cos 2\gamma = -1, \tag{10.19e}$$

which can be derived from (10.18) and (10.9) respectively. Finally Eq. (10.17) can be expressed as

$$\mu_r = \mu_r(\gamma, \dot{\phi}, \dot{\theta}_1, \dot{\theta}_0, A). \tag{10.19f}$$

The system (10.19) consists of six equations in six unknowns and is thus fully determined.

The energy shift of the CGS can be expressed by inserting the fields defined in Eq. (10.19) in the Hamiltonian, sum of H_{matt} (Eq. (2.36)) and $H_{em} = \frac{1}{2}\int_V(\vec{E}^2 + \vec{B}^2)d^3x$:

$$\frac{\delta E}{N} = E_{GS} - E_0 = \frac{\omega_0}{2}[2A^2(1 + 2\mu_r) + 3\,g\,\delta\,A\sin 2\gamma - \cos^2\gamma + 1]$$

$$= \omega_0[A^2(1 + 2\mu_r) + \sin^2\gamma + \frac{3}{2}\delta g\,A\sin 2\gamma]. \tag{10.20}$$

It is immediately checked that a negative $\delta E/N$ is only possible when $\delta = e^{i\chi} = -1$. Setting $\delta = -1$ from Eq. (10.19c) we get

$$\epsilon = \dot{\phi} - 1 = \sqrt{1 + 2\mu_r - \frac{g}{A}\sin 2\gamma}, \tag{10.21}$$

i.e.

$$A = \frac{g\sin 2\gamma}{1 + 2\mu_r - \epsilon^2}. \tag{10.22}$$

From eqs.(10.19a), (10.19b) and (10.19d) we easily obtain:

$$\epsilon = -1 + 2g\,A\cot\gamma, \tag{10.23}$$

which together with Eq. (10.19e) yields

$$\cos 2\gamma = -\frac{\epsilon\,(1 + \epsilon)}{2\,\epsilon^2 + \epsilon - (1 + 2\mu_r)}. \tag{10.24}$$

From Eqs.(10.22), (10.23) and (10.24) one can derive the following quartic equation:

$$2 \in^4 + \in^3 - 3 \in^2 (1 + 2\mu_r) - \in (2g^2 + 1 + 2\mu_r) + (1 + 2\mu_r)^2 = 0, \qquad (10.25)$$

which yields a positive solution for $\in$, as implied by Eq. (10.21), provided

$$1 + 2\mu_r + 2g^2 > 0. \qquad (10.26)$$

In order to evaluate g^2 one should know the density of the coherent liquid in its ground state. It would appear that according to Eq. (10.20) the system finds it energetically favourable to increase indefinitely its density, but as a matter of fact this conclusion is fallacious for in our analysis we have neglected the short-range repulsion that prevents such indefinite condensation to occur. In any case one should expect the liquid to finally reach a configuration where the molecules are packed as closely as possible. The fact that the intermolecular distance of liquid water ($\simeq 3.1\overset{\circ}{A}$) is rather larger than twice the molecular radius ($\simeq 1\overset{\circ}{A}$) in the ground state indicates that the (average) size of the molecule in the CGS is definitely larger than in the PGS (*i.e.* in the vapor). This is precisely what we expect, since the excited state at 12.06 eV from spectroscopic data[11,12] appears to be more extended than the ground state.

Since temperature effects will be discussed in the next Section (our analysis so far has been at $T = 0$), the expected density of the liquid at such temperature (should it exist) should be the limiting density that water reaches through supercooling, namely the ice relative density — 0.92 — that is measured at the lowest supercooling temperature reached so far ($\sim -45°C$)[8]. We will return to the problem of water density in the next Section.

The value of g^2 corresponding to a relative density 0.92 is 0.036 ± 0.003. Its small value allows us to simplify Eq. (10.13) as

$$g^2 > g_c^2 = \frac{1 + 2\mu_r}{2}, \qquad (10.27)$$

giving a very stringent bound for μ_r namely

$$-0.54 \leq \mu_r \leq -0.46. \qquad (10.28)$$

When μ_r is restricted as in (10.28), Eq. (10.25) admits a very small solution ($\in \simeq 10^{-2}$) corresponding to a negative $\frac{\delta E}{N}$ (Eq. (10.20)), and $\omega_r = \in \omega_0$ is of the order of a few tenths of eV. Consequently in the expression of μ_r, given in Eq. (10.17), in the rhs we need only keep the last term. We have:

$$\mu_r \simeq -\frac{3}{2} \left(\frac{\omega_p}{\omega_0}\right)^2 \frac{\sin^2 \gamma \, f_{01} \, \omega_r^2}{(E_0 - E_{GS})^2 - \omega_r^2} = -\frac{9}{4} \pi \frac{\sin^2 \gamma \, g^2 \omega_r^2}{(\frac{\delta E}{N})^2 - \omega_r^2}. \qquad (10.29)$$

Since $\mu_r \simeq -1/2$ and $\sin^2 \gamma g^2 \omega_r^2$ is a very small number we get a consistent solution only if

$$|\frac{\delta E}{N}| \simeq \omega_r = \omega_0 \in .$$ (10.30)

The complete solution of the system (10.19) thus yields

$$\frac{\delta E}{N} \simeq -0.26\text{eV},$$

$$A \simeq 2.40,$$

$$\mu_r = -0.4734,$$ (10.31)

$$\sin^2 \gamma = 0.127.$$

It is interesting that the small negative value of $\frac{\delta E}{N}$ arises, according to Eq. (10.20), from the compensation of the positive terms $\omega_0(1+2\mu_r)A^2 \simeq 3.69$ eV and $\omega_0 \sin^2 \gamma = 1.53$ eV needed to excite the e.m. field and the water molecules respectively by the negative interaction energy between the coherent matter and e.m. fields, which is $-\frac{3}{2}\omega_0 gA \sin 2\gamma = -5.48$ eV.

Let us now discuss the space dependence of the solutions of Eq. (10.19). From Chapter 2 we know that the e.m. field A has a gentle space dependence over distances of the order of $\lambda = \frac{2\pi}{\omega_0}$ (Coherence Domain) varying as $j_0(\omega_0 r) = \frac{\sin \omega_0 r}{\omega_0 r}$. Since $\in$ and μ_r do not vary within the CD[b], we note that according to Eq. (10.22) $\sin 2\gamma$ has the same spatial behaviour as A.

10.3   The thermodynamics of liquid water

We are now in a position to begin our investigation of the properties of our system, liquid water, in the CGS. A first, relevant observation is that we must expect that in the new ground state the short-range interactions too are affected, for we have just seen that in the CGS a water molecule is described by the superposition of two different states of the isolated molecule — the ground state and a particular excited state — whose relative probabilities, according to Eq. (10.31), are about 0.87 and 0.13. Thus all the physical properties of this new molecular state, such as the mean square radius, the electric dipole or the Lennard-Jones pair-potential etc., are expected to differ from those of the free molecule. In particular the "energy gap"

---

[b]The fact that μ_r does not vary within a CD stems from the fact that it is not a strictly local quantity and represents the "average photon mass" in the CD. In particular μ_r retains its value Eq. (10.31) anywhere in the CD.

between a molecule belonging to the "coherent liquid" (the CGS) and one of the "normal liquid", which finds itself in the perturbative ground state (*i.e.* at zero e.m. field), will acquire an extra electrostatic contribution, that will be denoted δ_{es}.

Next, let us bring temperature into the picture. As we have argued time and again, at $T \neq 0$ thermal fluctuations excite out of the CGS an "incoherent" fraction $F_n(T)$ of the total population of molecules. Such a fraction populates the states of a normal liquid following a Boltzmann distribution whose structure is determined by the single particle energy levels (rotons?) of the normal liquid. From Eq. (10.30) the thermal fluctuations can excite the CGS molecules more easily at the boundaries of the "coherence domains": starting from $x = \frac{3}{4}$ and proceeding inward towards $x = 0$ with increasing temperature.

At the temperatures where liquid water exists, the water molecules are usually assumed[13] to be unable to perform full rotations; only small librations around a fixed direction (hindered rotations) being accessible to them. The "energy gap" which protects the coherent phase against thermal fluctuations can then be written as a sum of two contributions:

$$\Delta(x) = \delta_c g(x) + \delta_{es}. \tag{10.32}$$

The next problem we must address is the "excitation curve" $\omega(k)$ of the single-particle levels of liquid water in the normal phase. In order to obtain a rather reliable excitation curve[c] we shall proceed as follows: for "low" momenta k $(k < 1\text{Å}^{-1})$[d] the excitation curve has a phononic structure:

$$\omega(k) \cong v_s k \quad (k < 1 \text{ Å}^{-1}), \tag{10.33}$$

where v_s is the velocity of sound $(v_s = 4 \cdot 10^{-6})$. For "high" k $(k > 9\text{Å}^{-1})$ the spectrum of excitation becomes the free water molecule spectrum:[e]

$$\omega(k) \simeq \frac{k^2}{2m} \quad (k > 9 \text{ Å}^{-1}). \tag{10.34}$$

---

[c]In the case of ^{4}He such curve was derived on purely thermodynamical arguments by Landau[14], and later confirmed by neutron scattering experiments[15]. Unfortunately the direct determination is impossible for water (at the temperatures where it exists, water is **not** superfluid), so we shall follow a course that is similar to Landau's.

[d]The choice of the value $k = 1\text{Å}^{-1}$ for the highest momentum of the phononic region stems from the observation that "quasi-particles" with such momentum represent still a collective excitation of the liquid, as evidenced, for instance, in liquid ^{4}He.

[e]The space-resolution $\frac{\pi}{k}$ of a quasi-particle for $k \geq 9\text{Å}^{-1}$ is better than 0.35Å, and at this distance the quasi-particle is certainly well described by an isolated molecule, hence the usual free, non-relativistic dispersion relation.

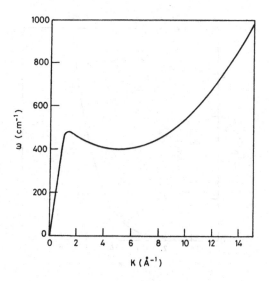

Fig. 10.2. The excitation curve $\omega = \omega(k)$ of the thermal excitations which comprise the "normal phase" (k in $\overset{\circ}{A}^{-1}$)

As a reasonable interpolation, in the region ($1\overset{\circ}{A}^{-1} < k < 9\overset{\circ}{A}^{-1}$) following Landau's Ansatz for liquid ^{4}He, we assume

$$\omega(k) = \delta_0 + \frac{(k - k_0)^2}{2m},\qquad(10.35)$$

where a smooth connection with Eq. (10.33) and Eq. (10.34) fixes

$$
\begin{aligned}
k_0 &\simeq 5\overset{\circ}{A}^{-1}\\
\delta_0 &\simeq 400\ \text{cm}^{-1}.
\end{aligned}\qquad(10.36)
$$

The full excitation curve we shall use is depicted in Fig. 10.2.[f] It is now straightforward (again following Landau) to get for the partition function:

$$Z = \left(\frac{V}{N}\right)\frac{k_0^2}{2\pi}\left(\frac{mT}{2\pi}\right)^{\frac{1}{2}} e^{-\delta_0/T},\qquad(10.37)$$

and, for the space-dependent "incoherent" fraction ($\delta_c = 0.26$ eV, see Eq. (10.30))

---

[f]Note that in simple liquids there is a quite compelling evidence for excitation curves of such structure.

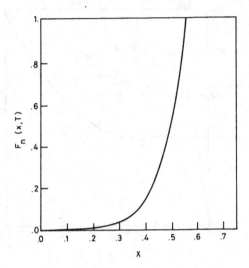

Fig. 10.3. The "local" normal phase fraction $F_n(x, T)$ plotted against x at $T = 300$ K.

$$F_n(x, T) = Z e^{\frac{-\Delta(x)}{T}}$$
$$\Delta(x) = \delta_c g(x) + \delta_{es} \tag{10.38}$$

from which we derive the "coherent fraction" $F_c(T)$

$$F_c(T) = 1 - \frac{4\pi}{\frac{4\pi}{3}\left(\frac{3}{4}\right)^3} \int_0^{\frac{3}{4}} dx \; x^2 F_n(x, t). \tag{10.39}$$

In Fig. 10.3 we report the space dependence of $F_n(x, T)$ at $T = 300$ K and $\delta_{es} = 0.022$ eV. The very steep rise of $F_n(x, T)$ from zero to one allows us to give a simple geometric meaning to $F_c(T)$. Indeed for $F_c(T)$ we can take the fraction of molecules contained in the sphere of radius $\bar{x}(T)$,

$$F_c(T) = \frac{64}{27} \bar{x}(T)^3. \tag{10.40}$$

Fig. 10.4 shows $F_c(T)$ and Fig. 10.5 its derivative $\frac{dF_c}{dT}$. We wish to note here that, unfortunately, we are not yet in a position to **theoretically determine** $F_c(T)$, since we need to know the "static energy gap" δ_{es} in Eq. (10.37) – (10.38), which requires

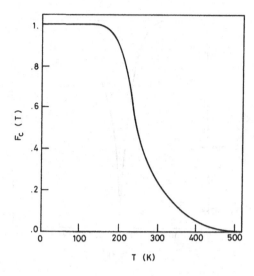

Fig. 10.4. The "coherent fraction" $F_c(T)$ vs. T

in turn that we know in the CGS the values of the density $\left(\frac{N}{V}\right)_c$, of the electric dipole D_c, and also the structure of the dispersive forces, that as we have argued, demands a molecular calculation that has not yet been performed. However, the shape of $\frac{dF_c}{dT}$ (Fig. 10.5) is typical and shows a very narrow negative peak, where the derivatives of thermodynamical quantities, such as specific heat and compressibility, should exhibit a steep increase. Actually this behaviour occurs at about 230 K[8], and having been able to predict the occurrence of this kind of behaviour should be looked upon as a success of our approach. In order to have the narrow negative peak just at this temperature we need choose $\delta_{es} = 0.022$ eV. Note that δ_{es} depends in general on T.

The dependence of δ_{es} upon T stems from the (linear) dependence of the short-range electrostatic energy on density, which in turn depends on T. Indeed we have shown in Section (1.3) that the energy associated with a pair-interaction potential $V_p(\vec{x})$ for an ensemble of molecules described by the quantum field $\psi(\vec{x})$ is given simply by

$$U_p = \frac{1}{2} \int_V d^3x \int_V d^3y\, \psi^+(\vec{x})\psi(\vec{x}) V_p(\vec{x} - \vec{y})\psi^+(\vec{y})\psi(\vec{y}), \qquad (10.41)$$

which in the thermodynamic limit ($N \to \infty$, $\frac{N}{V}$ fixed), and in the "mean field" approximation $|\psi(x)|^2 \simeq \left(\frac{N}{V}\right)$ yields:

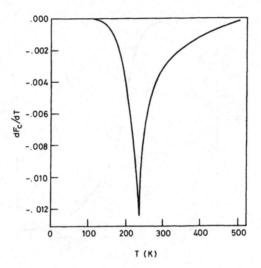

Fig. 10.5. The temperature derivative $\frac{dF_c(T)}{dT}$ vs. T.

$$\frac{U_p}{N} = \frac{1}{2}\left(\frac{N}{V}\right)\int V_p(\vec{x})d^3x = \frac{1}{2}\left(\frac{N}{V}\right)V_0. \qquad (10.42)$$

Thus we may write for the energy originating from the short-range interactions

$$\frac{U_{sr}}{N} = \delta_{sr}\rho(T), \qquad (10.43)$$

where $\rho(T)$ is the density of water relative to its maximum density, that occurs at $T = 4°C$.

Note that, since we have assumed that the density of the coherent phase is $\rho_c = 0.92$ (independent of temperature), we may evidently express the observed density of water as:

$$\rho(T) = 0.92F_c(T) + (1 - F_c(T))\rho_n(T), \qquad (10.44)$$

where the "normal phase" density $\rho_n(T)$ can be expanded around the temperature T_0 of maximum density ($T_0 = 277$ K) as

$$\rho_n(T) = \rho_n(T_0) - a(T - T_0). \qquad (10.45)$$

We can now fix $\rho_n(T_0)$ and a by the following two requirements:

$$\rho(T_0) = 1 = 0.92 F_c(T_0) + (1 - F_c(T_0))\rho_n(T_0), \qquad (10.46)$$

and

$$\frac{d\rho(T_0)}{dT} = 0 = -\frac{dF_c}{dT}(T_0)(0.92 - \rho_n(T_0)) - a(1 - F_c(T_0)), \qquad (10.47)$$

expressing the normalization $\rho(T_0) = 1$ of the density of water, and the peculiar feature of water of having its maximum density at $T = T_0 = 277\ K$.

Thus we obtain:

$$\rho_n(T_0) = \frac{1 - 0.92 F_c(T_0)}{1 - F_c(T_0)} \qquad (10.48a)$$

and

$$a = -\frac{0.08}{(1 - F_c(T))^2} \frac{dF_c}{dT}(T_0). \qquad (10.48b)$$

Eq. (10.44), supplemented by the explicit expressions (10.48), explains (qualitatively) in a simple and elegant fashion the celebrated density anomaly of water. Indeed at $T < 230K$, in the supercooled water $F_c(T_S) \geq 0.8$ and the density of water is near 0.92; by increasing the temperature the normal phase gets formed, at the interstices between coherent domains, with the higher density which, however, naturally decreases with temperature (see Eq. (10.45) and (10.48b) expressing the positivity of a): the density maximum occurs at a point where the increase of the fraction $1 - F_c$ of the normal phase of higher density is just compensated by the decrease of the normal phase density with temperature. A similar behaviour is expected for the isothermal compressibility $\kappa_T = \frac{\partial \log V}{\partial p}|_T$, once the relationship $\bar{x} = \bar{x}(p)$ is known. $\rho(T)$ is plotted in Fig. 10.6, where it is compared with the experimental curve.

We now consider the specific heat C_p at constant pressure. When the temperature T is increased by the infinitesimal quantity dT a fraction $dF_c(T)$ of the coherent phase "evaporates" from the surface of the coherent phase into the normal phase. In order to do this it must overcome an energy gap $\Delta[\bar{x}(T)] = \delta_c(0)g[\bar{x}(T)] + \delta_{es}$, finally reaching the normal phase whose enthalpy is

$$\begin{aligned} H &= \delta_{SR}^{(n)}\rho_n(T) + \delta_0 + T\frac{\partial}{\partial T}\log Z \\ &= \delta_{SR}^{(n)}\rho_n(T) + \delta_0 + \frac{T}{2}. \end{aligned} \qquad (10.49)$$

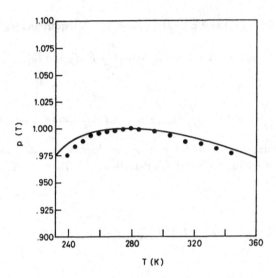

Fig. 10.6. The calculated curve of the relative density $\rho = \rho(T)$ (Eq. (10.44)) compared with the experimental points (Ref. [7]).

Thus the specific heat C_p will comprise two contributions[g]

$$C_p = \frac{dF_c}{dT}\left[\delta_c(0)g[\bar{x}(T)] + \delta_{es} + \frac{T}{2} + \delta_0\right] + (1 - F_c)\left[a\delta_{SR}^{(n)} + \frac{1}{2}\right], \qquad (10.50)$$

the first expressing the energy needed to transfer the molecule from the coherent phase to the average energy of the normal phase at the given temperature, while the second is just the C_p of the normal phase. Setting $\delta_{SR}^{(n)} = 0.3$ eV (a value suggested, as we shall see in a moment, by the phenomenology of the van der Waals' equation), C_p as a function of T is plotted in Fig. 10.7, where it is compared with the experimental curve. As one can see the agreement is reasonable and of the same quality of the determination of the slope a of $\rho(T)$ from $F_c(T)$. We must note that the contribution of the normal phase to Eq. (10.50) is underestimated since we have totally neglected rotations, whose contribution could become significant at higher temperatures[17].

Let us conclude this Section by a discussion of the liquid-vapor phase transition. One observes that in liquid water the "normal phase" is prevented from escaping into the gas phase, thus "sticking" to the "coherent phase", by the "chemical potential" $F_c[\delta_{cg}(x) + \delta_{es}]$, which prevails over the "entropic gain" $T\Delta S$ that could be

---

[g]The reader should recall that throughout this book we are using the "natural" units in which $k_B = 1$. This means that for $T = 1$ K, $k_B T = 4.36$ cm$^{-1} = 8.7\ 10^{-5}$ eV.

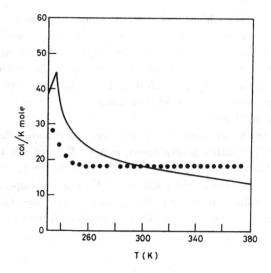

Fig. 10.7. The specific heat C_p plotted vs. T, compared with the experimental points (Ref. [7]).

achieved by the water molecules leaving the liquid and evaporating to the gas phase. Quantitatively, the Gibbs potential of the "normal phase" of the liquid is given by

$$G_L^{(n)} = F_c[-\delta_c \bar{g}(x) - \delta_{es} - \delta_{SR}\rho_c]$$
$$+ (1 - F_c)\left[-\delta_{SR}\rho_n + \frac{T}{2} + \delta_0 + pV_L - TS_L\right], \tag{10.51}$$

where $\bar{g}(x)$ is the average of $g(x)$ over the sphere of radius $\bar{x}(T)$, and S_L is the entropy of the "gas of rotons" whose spectrum is given by Eq. (10.35).

The Gibbs potential for the same ensemble of molecules in the vapor phase is:

$$G_V^{(n)} = (1 - F_C)[3T + pV_V - TS_V], \tag{10.52}$$

and the phase transition is expected to occur when $G_L^{(n)} = G_V^{(n)}$, i.e. at the temperature T_b where

$$\frac{F_c(T_b)}{1 - F_c(T_b)} = \frac{T_b(S_V - S_L - 7/2) + \delta_0 - \delta_{SR}\rho_n(T_b)}{\delta_c \bar{g}(x) + \delta_{es} + 0.92\delta_{SR}}, \tag{10.53}$$

where we have set, with good accuracy, $p(V_V - V_L) = T$.

When T reaches T_b each additional external supply of energy cannot heat the system, which maintains the two Gibbs potentials equal by transferring molecules from the "normal phase" of the liquid to their vapor phase and replacing them by other molecules that get extracted from the "reservoir" of the coherent phase. Such a frantic molecular traffic, and the typical bubbling that accompanies it, lasts until the "reservoir" gets completely depleted: all molecules are now in the vapor phase and T can resume its increase.

Getting now to the numbers, as we have mentioned above, the only quantity that in Eq. (10.54) is rather poorly known is δ_{SR}. It turns out that in order to get consistently the observed $T_b = 373$ K one must choose $\delta_{SR} = 0.3$eV, which would correspond to a second virial coefficient at the critical temperature $B^{(2)}(T_c) = 4.4$ l^2atm/mole^2, which is only about 20% lower than experimentally observed. As for the entropy S_V appearing in Eq. (10.54), we evaluate it from the vapor partition function:

$$Z_V = Z_V^{(tr)} Z_V^{(rot)} = \frac{V_V}{\pi^{5/2}} T^3 (mI_1)^{\frac{1}{2}} (mI_2)^{\frac{1}{2}} (mI_3)^{\frac{1}{2}}, \qquad (10.54)$$

where I_1, I_2 and I_3 are the moments of inertia of the water molecules with respect to the principal axes. From spectroscopical studies one gets[16] for $2\pi I_i (i = 1, 2, 3)$ the values 23.79 cm^{-1}, 37.14 cm^{-1} and 42.36 cm^{-1} respectively. As for S_L it is computed from the partition function given in Eq. (10.37). For the entropy difference one gets the simple expression

$$S_V - S_L = \log\left(\frac{Z_V}{Z_L}\right) = \log\left[\frac{V_V}{V_L} \frac{mT_b^{5/2}}{2\sqrt{2}\pi} \frac{(I_1 I_2 I_3)^{\frac{1}{2}}}{k_0^2}\right], \qquad (10.55)$$

and putting numbers in one computes for $T_b = 373$K

$$T_b(S_V - S_L) = 0.33 \text{ eV}. \qquad (10.56)$$

We can also determine numerically the latent heat of vaporization from the relation:

$$E/N = F_c(T_b)[\delta_c(0)\bar{g}(x) + \delta_{es} + \frac{T}{2} + \delta_0] + (\delta_{SR} + \frac{7}{2}T), \qquad (10.57)$$

whose interpretation is simple: it is just the energy per molecule needed to evaporate the liquid phase and to liberate the coherent phase whose fraction is precisely $F_c(T_b)$.

Recalling that $F_c(T_b) = 0.08$, we easily compute for the latent heat

$$L \cong 9.8 \text{ Kcal/mole}, \qquad (10.58)$$

to be compared with the observed 9.7 Kcal/mole.

10.4 Outlook

I wish to conclude this Chapter by recalling first the main features of the dynamical scenario that QED coherence paints of liquid water:

(i) liquid water has been found to consist, as all QED coherent condensed matter systems at non-zero temperature, of two distinct interspersed phases;

(ii) the incoherent phase comprises water molecules in the molecular ground state (as observed in the gas phase) packed in a highly dense state in the interstices around large clusters (of the size of 750 $\mathring{A}$) in which the water molecules interact coherently with a large classical electromagnetic field;

(iii) the coherent phase is just built up by the collection of the "coherent domains" that at a given temperature have survived the disordering attacks of thermal fluctuations. In the coherent phase, due to the oscillations of the water molecules between the ground state and an excited state at 12.06 eV (see Section (10.2)), the volume occupied by each molecule turns out to be definitely larger than the volume occupied by the molecules of the incoherent phase, thus rendering the coherent phase density much closer to the density of ice;

(iv) the energy gap $\delta_c g(x)$ (see Section (10.3)) that protects the molecules from "evaporating" from the coherent phase into the incoherent, gaseous one, allows us to evaluate the relative populations $F_c(T)[1 - F_c(T)]$ of the coherent (incoherent) molecules as a function of T, by assuming, à la Landau, a roton-like excitation phase of liquid water;

(v) by representing the expected difference in the electrostatic energy between the coherent and incoherent phases with a small and negative quantity $\delta_{es} \simeq -0.022$ eV, we are able to predict the existence of a temperature interval around 230K, where $|\frac{dF_c}{dT}|$ exhibits a sharp peak. This striking deduction from our model (see Fig. 10.5) can easily account for the well documented behaviours[8] of the specific heat and compressibility of supercooled water at $T = 230$ K;

(vi) due to the complete description of the dynamics of the coherent phase and of the thermodynamics of the incoherent phase (through the excitation spectrum, and the electrostatic energy as inferred from the experimentally observed second virial coefficient), and the knowledge of $F_c(T)$ (see Fig. 10.4), we can compute the specific heat of liquid water (see Fig. 10.7) and, within the scope of our approximations, we obtain a rather pleasing agreement with experiments. As a by-product we can also understand **qualitatively** the density anomaly of water as due to the superposition with temperature dependent weights of two different temperature dependent densities;

(vii) as for the phase diagram we now have an understanding of the critical volume, V_{crit} (see Eq. (10.13)), as the largest molar volume above which no QED coherent process can spontaneously take place. As for the boiling temperature, T_b, we have derived it, in agreement with experiment, from the analysis of the Gibbs potentials of water in the liquid and in the gas phase, and obtained an interesting physical picture of the boiling process;

(viii) the latent heat of vaporization has also been determined from our dynamical analysis, and found in good agreement with the measured value (see Eq. (10.58)).

Besides the obvious encouragement that we draw from the satisfactory agreement between the theoretical and the experimental thermodynamical quantities, the aspect of the above deductions that we find more intellectually pleasing is that the extremely ingenious research program initiated by Linus Pauling, based on the idea of a "flickering" H-bonded network of the molecules of water, turns out to find (with some differences, however) its theoretical foundation. Indeed the clusters of H-bonded molecules that, due to their open structure, occupy larger volumes are, in our view, nothing but the coherence domains, where molecules evolve in phase with a coherent e.m. field (as explained in Section (10.2)). And once this identification is made, there is no big difficulty in seeing that the logical steps one performs in the two different research programs are just (essentially) the same.

It should, however, be stressed that one unsatisfactory aspect of the H-bonded network program, that of its "flickering" nature, is in our approach replaced by a completely **ordered, stable** dynamical structure, the coherence domains of water molecules oscillating in phase with a large, classical e.m. field. Once we appreciate the considerable amount of coherence that characterizes our theory of liquid water, it is not impossible to imagine that such marvelously ordered structure may retain and release electromagnetic information that it has acquired in some way or other. This remark is only intended to warn against considering water as just a large collection of more or less insignificant, little molecules.

References to Chapter 10

1. R. Arani, I. Bono, E. Del Giudice and G. Preparata, Preprint MITH 93/3 of Physics Department of University of Milano (Italia).

2. F. Franks (ed), *Water: a comprehensive treatise* (Plenum Press, New York, 1972-1982).

3. W.K. Röntgen, *Ann. Phys.* **45** (1892) 91.

4. J.D. Bernal and R.H. Fowler, *J. Chem. Phys.* **1** (1933) 515.

5. G. Nemethy and H.A. Scheraga, *J. Chem. Phys.* **36** (1962) 3382.

6. H.E. Stanley and J. Teixeira, *J. Chem. Phys.* **73** (1980) 3404.

7. D.Bertolini, M. Cassettari, M. Ferrario, P. Grigolini, G. Salvetti and A. Tani, *J.Chem.Phys.* **91** (1989) 1179.

8. C.A. Angell, *Supercooled water*, Chapter 1 of Vol. 7 of Ref. 1.

9. A. Rahman and F.H. Stillinger, *J. Am. Chem. Soc.* **95** (1973) 7943.

10. A.D. Buckingham, in *Water and Aqueous Solution*, eds. G.W. Neilson and J.E. Enderly (Colston Papers no. 37, Hilger, Bristol and Boston, 1985); see in particular the discussion on pages 11, 12 and 13.

11. P. Gurtler, V. Saile and E.E. Koch, *Chem. Phys. Lett.* **51** (1977) 2.

12. G. Herzberg, *Electronic Spectra of Polyatomic Molecules* (Van Nostrand Reinhold Company, 1966).

13. L. Pauling, *General Chemistry*, (Dover, New York, 1982).

14. L. Landau, *J. Phys. USSR* **5** (1941) 5; it J.Phys.USSR **11** (1947) 91.

15. A.D.B. Woods and E.C. Svensson, *Phys. Rev. Lett.* **41** (1978) 974.

16. C.W. Kern and M. Karplus, *The Water Molecule* Chapt. 2 of Vol. 1 of Ref. 1.

17. G.S. Kell, *Thermodynamic and Transport Properties of Fluid Water* Chapt. 10 of Vol. 1 of Ref. 1.

Chapter 11

A FAR REACHING ANALOGY: QCD COHERENCE IN NUCLEAR MATTER

11.1 Pions, nucleons and Δ-resonances: the building blocks of Nuclei and Nuclear Matter

In this last Chapter our attention will be devoted to the extension of the main ideas of QED coherence to a physical system, Nuclear Matter, whose dynamics is not governed by QED, but instead by Quantum Chromo Dynamics (QCD).

Up until the developments that are the subject of this Chapter the commonly held view of Nuclei and Nuclear Matter has been very close to the prevailing view of ordinary condensed matter, according to which its elementary constituents are kept together by an electrostatic, hence short-ranged glue. Thus in "orthodox" Nuclear Physics (NP) nuclei from the lightest (the deuteron) to the heaviest transuranic elements are all believed to arise from the short-range "nucleostatic" forces that are generated by the exchange of low-mass mesons (π, ρ, ω, η, etc.) between nucleons, the isospin doublets comprising the proton ($I_3 = \frac{1}{2}$) and the neutron ($I_3 = -\frac{1}{2}$). The rather large masses of the exchanged mesons produce Yukawa-like forces whose ranges are of the order of 1 F or less.[a] If we now scale space (and time) by a factor of about 10^5, the analogy with the Debye screened electrostatic interaction of ordinary condensed matter becomes quite punctual, for here too we have different types of "hooks", of the static variety, the longest ranged being due to π-meson exchange. In particular the universally attractive force, that is believed to be fundamental for the existence of large nuclei, arising from the exchange of two-π's , is just the analogue of the van der Waals interaction, the second order exchange of the photons being replaced by the second order exchange of pions. Its range, however, is quite small,

---

[a]Recall that the Compton wave-length of the lightest meson, the $\pi(140)$, is $\frac{1}{m_\pi} \simeq 1.4 \ 10^{-13}$ cm = 1.4 F

about $\frac{1}{2m_\pi} = 0.7$ F, and comparable with the size of the nucleon (N) whose root-mean square radius is of the order of 0.8 F.

What kind of structure would one expect for systems consisting of a few nucleons, bound by forces of this kind? From the analogy with the electrostatically glued ordinary matter, the answer would seem rather straightforward: forces (the two-π exchange), whose range is shorter than the size of the bound particles (the nucleon), should give rise to clusters whose constituents oscillate about their equilibrium positions with oscillation amplitudes of the order of 1 F and frequencies of the order of the binding energy, about 10 MeV.

As we know well, the Nuclei, even the light ones, have nothing to do with such configurations! The highly successful Shell Model (SM) simply reminds us that the energy levels of N's in a Nucleus are arranged in shells, that group the almost degenerate single particle levels of N's moving in a central potential. Rather than a "democratic" cluster of nucleons held together by the short-range "hooks" of the two-π exchange, the picture that the SM evokes is that of the atomic electrons that are neatly arranged in the shells of the Mendeleef table and incessantly revolve around their "sun", the Nucleus. Thus if one legitimately wanted to give a solid, realistic motivation for this analogy, one would be hard pressed to understand what in the Nucleus, that in terms of its identical fermionic N's looks so "democratic", may play the rôle of the "sun", arranging the single nucleon orbits in the strict hierarchy of a rather simple (Wood-Saxon, Reid soft-core etc.) central potential. It seems undeniable that if the main actors of the dynamics of a Nucleus are the nucleons, and the forces are the short-ranged two-body forces generated by low-mass meson exchange, the "democratic" cluster picture is unavoidable. Logics and the Ockham's razor allow no alternative: if no hierarchy is present to start with, none is to be expected to arise in the course of the evolution, for this would require a fundamental dynamical rearrangement of the ground state, a kind of phase transition, which for a system of a few nucleons, bound by extremely short-ranged forces, is definitely excluded.

It is interesting, however, that confronted with a major theoretical difficulty the "orthodox" nuclear physicist has reacted just in the same way as the "orthodox" condensed matter physicist when confronted by the strange behaviour of electrons in a metal: he has lowered his aim from explanation to consistency. According to him, in fact, there is no need to explain why all this happens (for it clearly happens), what is necessary to show is only that if one assumes the existence of the appropriate central potential, then the N's motions in the single particle orbits generate a potential that is consistent with the assumed one. The circle thus closes but what clearly is left out is just the fundamental dynamical reason **why** those set of forces, acting on that set of nucleons, manage to produce such a beautifully ordered structure as a Nucleus. The natural philosopher has more than one reason to be unhappy: something seems definitely missing here.

Suppose now that we too were to push forward the analogy between NP and CMP, but this time in the area of coherent dynamical behaviour: what would we need to do? Evidently we should identify a nuclear matter-wave field, endowed with a set of frequencies, and a bosonic quantum field coupled to the matter field, inducing transitions between the different levels of the matter field. Assuming that this can be accomplished, and that the study of the dynamical evolution of the system matter-field produces stationary, and stable coherent configurations, would we understand the observed structure of the Nucleus? The answer is clearly yes, for we would be in the situation, already studied in Chapter 4, where the QED coherence of a system of two-level fermions coupled to the transverse e.m. field was studied and applied to liquid ^{3}He. We recall from Section (4.4) that, once involved in a coherent oscillation in phase with the e.m. field, the elementary fermions arrange themselves in the available (according to the Pauli principle) single particle states, belonging to a single particle Hamiltonian, with a potential term whose range spans the whole system and the (average) depth is given by the gap (binding energy per particle) generated by the coherent process: just what is needed to explain the fantomatic central potential of the SM! Thus looked from the side of fermionic nuclear matter (the nucleons), the hierarchical element necessary to produce the SM turns out to be just the coherent dynamics of the system matter-field.

Now that we know that a coherent dynamics of the nuclear constituents suffices, and indeed appears to be required **to explain** the remarkable properties of the Nucleus, we must try to see whether and in what way the familiar conditions for coherence are met. Let us consider first the nuclear matter wave-field:[b] the Quark Model (QM) (out of which the fundamental QCD-theory grew at the beginning of the seventies) tells us that the nucleon is a composite of three (spin $\frac{1}{2}$) quarks of the u (charge $\frac{2}{3}$) and d (charge $-\frac{1}{3}$) type, thus the proton is a uud- while the neutron is a udd-composite. One of the great successes of the QM is that it gives a simple explanation of the existence of a very crowded zoo of so called "elementary" particles, the majority of which were discovered in the fifties and sixties at the large accelerators of high-energy physics. In particular the QM shows that the nucleon (N(940)) and the Δ-resonance ($\Delta(1232)$) belong to the same multiplet, having the same spatial wave-function (S-wave) and differing only for the total spin-configuration: $S = \frac{1}{2}$ for N and $S = \frac{3}{2}$ for Δ. Thus the comparatively small width ($\Gamma_\Delta = 120$ MeV) of the Δ and its rather close mass ($m_\Delta = 1232$ MeV) to the mass of the N ($m_N = 940$ MeV) make it an ideal partner of the nucleon in a possible coherent dynamical evolution inside the Nucleus. In this way we are naturally led to introduce the nuclear matter

---

[b]This discussion follows closely the work in Ref. [1]

wave-field

$$\psi(\vec{x}, \alpha, t) = \psi_N(\vec{x}, t) + \psi_\Delta(\vec{x}, t), \tag{11.1}$$

where

$$\psi_N(\vec{x}, t) = \sum_{st} N_{st}(\vec{x}, t) \chi_s \eta_t \tag{11.2a}$$

$\chi_s(s = \pm\frac{1}{2})$ is the spin-$\frac{1}{2}$ wave-function, while $\eta_t(t = \pm\frac{1}{2})$ is the isospin-$\frac{1}{2}$ wave-function, and

$$\psi_\Delta(\vec{x}, t) = \sum_{ST} \Delta_{ST}(\vec{x}, t) \chi_S \eta_T, \tag{11.2b}$$

where $\chi_S(S = \pm\frac{3}{2}, \pm\frac{1}{2})$ and $\eta_T(T = \pm\frac{3}{2}, \pm\frac{1}{2})$ are the spin and isospin wave-functions of the Δ.

Now that we have identified the two levels of a possible coherent nuclear dynamics, the N and the Δ, and a discrete frequency $\omega = m_\Delta - m_N \cong 300$ MeV, we must find the nuclear field connecting the two states. The search is not difficult: the Δ-resonance owes its finite width to the decay $\Delta \to N\pi$; thus the nuclear field is readily identified with the π-field,

$$\Phi_k(\vec{x}, t) = \sum_{\vec{q}} \frac{1}{(2\omega_{\vec{q}} V)^{\frac{1}{2}}} \left[a_{k\vec{q}}(t) e^{-i(\omega_{\vec{q}} t - \vec{q}\vec{x})} + a_{k\vec{q}}^+(t)^{i(\omega_{\vec{q}} t - \vec{q}\vec{x})} \right]$$

where k= 1, 2, 3 is the isospin index, and

$$\omega_{\vec{q}} = \sqrt{\vec{q}^{\,2} + m_\pi^2}.$$

We are now ready to write down the nuclear Hamiltonian H_N, which can be split into four pieces:

$$H_N = H_K + H_{INT} + H_\pi + H_{SR}, \tag{11.3}$$

with the kinetic Hamiltonian

$$H_K = \int_V d^3x \left[\psi_N^+ \left(m_N - \frac{\vec{\nabla}^2}{2m_N} \right) \psi_N + \psi_\Delta^+ \left(m_\Delta - \frac{\vec{\nabla}^2}{2m_\Delta} \right) \psi_\Delta \right], \tag{11.4a}$$

the interaction Hamiltonian

$$H_{INT} = \frac{g}{(4m_\Delta m_N)^{\frac{1}{2}}} \left[\int_V d^3x \Delta_{ST}^+ N_{st} C_{Ss}^a C_{Tt}^k \vec{\nabla}_a \Phi_k + \text{h.c.} \right], \tag{11.4b}$$

where

$$C_{Ss}^a = \chi_S^{+ijk} \in_{ir} (\sigma^a)_j^r \chi_{s,k}.$$ (11.4c)

Note that, due to parity conservation, the interaction N-Δ-π is in P-wave: this explains the gradient $\vec{\nabla}\Phi_k$ appearing in H_{INT}. As for H_π one has in full analogy with the e.m. case:

$$H_\pi = \sum_{\vec{q}k} \frac{1}{2\omega_{\vec{q}}} \left[2\omega_{\vec{q}}^2 a_{k\vec{q}}^+ a_{k\vec{q}} + i\omega_{\vec{q}}(a_{k\vec{q}}^+ \dot{a}_{k\vec{q}} - \dot{a}_{k\vec{q}}^+ a_{k\vec{q}}) + \dot{a}_{k\vec{q}}^+ \dot{a}_{k\vec{q}} \right].$$ (11.4d)

Finally the short-range Hamiltonian, which can be schematically written as

$$H_{SR} = \frac{1}{2} \int_V d^3x d^3y \, \psi_N^+(\vec{x},t)\psi_N(\vec{x},t)V_{SR}(\vec{x}-\vec{y})\psi_N^+(\vec{y},t)\psi_N(\vec{y},t),$$ (11.4e)

the short range potential $V_{SR}(\vec{x})$ provides the "hard-core" that keeps the nuclear matter density $\frac{A}{V}$ (A is the atomic number, i.e. the number of nucleons of a nucleus) at its observed (almost) constant value. Apart from this rather relevant observation we shall have nothing to say about this piece of the Hamiltonian in the rest of this Chapter.

The coupling g, appearing in (11.4b), is completely determined by the observed width $\Gamma(\Delta \to N\pi)$, and one finds $\frac{g^2}{4\pi} \simeq 20$. All fundamental QCD inputs — masses and coupling constants — being available, we can now analyse the fundamental question of the possible coherent solutions of the dynamical question posed by the Hamiltonian H_N of Eq. (11.3).

This we shall do in the next Section.

11.2 The weak coherence of the $N\Delta\pi$ System: towards an understanding of the Nuclear Models

Our starting point is the Lagrangian:

$$
\begin{aligned}
L(t) = \int_V d^3x &\left\{ \psi_N^\dagger(\vec{x},t)i\frac{\partial}{\partial t}\psi_N(\vec{x},t) + \psi_\Delta^\dagger(\vec{x},t)i\frac{\partial}{\partial t}\psi_\Delta(\vec{x},t) \right\} \\
&+ \sum_{\vec{q}k} \left[\frac{1}{2}(a_{k\vec{q}}^+ \dot{a}_{k\vec{q}} - \dot{a}_{k\vec{q}}^+ a_{k\vec{q}}) + \frac{1}{2\omega_{\vec{q}}}\dot{a}_{k\vec{q}}^+ \dot{a}_{k\vec{q}} \right] - H_K - H_{INT},
\end{aligned}
$$ (11.5)

which, as usual, admits the conserved nucleon number operator

$$\hat{N} = \int_V d^3x \left[\psi_N^+(\vec{x},t)\psi_N(\vec{x},t) + \psi_\Delta^+(\vec{x},t)\psi_\Delta(\vec{x},t) \right] = A.$$

As we have seen in Chapter 2 and 3, the Euler-Lagrange equations following from (11.5) in their most general form are quite complicated, thus we shall consider them inside the coherence domains where the space-dependence can be neglected. In order to identify the radius R_{CD} of the coherence domains we observe that the frequency of the resonating mode, the only one that shall be retained, is

$$\omega_{\vec{q}} = \sqrt{\vec{q}^2 + m_\pi^2} = m_\Delta - m_N \simeq 300 \text{ MeV},$$

and that the already noted P-wave character of the interaction demands coherence over a region whose radius is given by the first zero of the spherical Bessel function $j_1(qr)$; thus for R_{CD} we may write

$$R_{CD} = \frac{4.49}{q} \simeq 3.5 \text{ F} \xrightarrow[\text{finite } \pi \text{ radius}]{} 4.2 \text{ F}, \tag{11.6}$$

where in the final estimate we have taken into account the well known fact that the pion has a finite radius (0.8 F). Thus taking from experiment $\frac{A}{V} = \left(\frac{4}{3}\pi R_0^3\right)^{-1}$ (with $R_0 \simeq 1.3$ F) we estimate that a coherence domain contains about 70 N/Δ which, as a result, will be described by single wave-functions that oscillate in phase with the "resonating" modes of the π-field.

Setting,

$$N_{st} = \sqrt{\frac{A}{V}} \, n_{st} e^{-im_N t}, \tag{11.7a}$$

$$\Delta_{ST} = \sqrt{\frac{A}{V}} \, \delta_{ST} e^{-im_\Delta t}, \tag{11.7b}$$

$$a_{k\vec{q}} = \sqrt{\frac{4\pi}{3} A} \, \alpha_{kj} \frac{q_j}{q}, \tag{11.7c}$$

the Euler-Lagrange equations within a coherence domain are ($\tau = \omega_{\vec{q}} \, t$)

$$\dot{n}_{st} = -G \, C_{sS}^i C_{tT}^b \alpha_{bi}^* \delta_{ST}, \tag{11.8a}$$

$$\dot{\delta}_{ST} = G \, C_{sS}^i C_{tT}^b \alpha_{bi} n_{st}, \tag{11.8b}$$

$$\dot{\alpha}_{bi} + \frac{i}{2}\ddot{\alpha}_{bi} = -G \, C_{sS}^i C_{tT}^b n_{st}^* \delta_{ST}, \tag{11.8c}$$

where

$$G = \sqrt{\frac{\pi}{24}}\, g\, \frac{q}{\sqrt{m_N m_\Delta}}\sqrt{\frac{A}{V}}\,\frac{1}{\omega_q^{\frac{3}{2}}}.\qquad(11.8d)$$

For short times the system (11.8) can be linearized to:

$$\ddot{\alpha}_{bi} + \frac{i}{2}\,\dot{\alpha}_{bi} + \frac{16}{9}\,G^2\alpha_{bi} = 0,\qquad(11.9)$$

with

$$\frac{16}{9}\,G^2 \cong 0.52 < g_c^2 = \frac{16}{27}.\qquad(11.10)$$

The effective coupling $\frac{16}{9}\,G^2$ being below the threshold value $\frac{16}{27}$, the system (11.8) does not run-away, and the field amplitudes retain the size of the vacuum fluctuations: we are in the regime of weak coherence already encountered in Chapter 4, in the case of the two isotopes of Helium.

Following the usual procedure we seek the stationary solutions of (11.8) of the form:

$$n_{st} = N_{st}e^{i\nu(t)},\quad N_{st}^*N_{st} = \cos^2\theta\qquad(11.11a)$$

$$\delta_{ST} = \Delta_{ST}e^{i\delta(t)},\quad \Delta_{ST}\Delta_{ST} = \sin^2\theta\qquad(11.11b)$$

$$\alpha_{bi} = a_{bi}e^{i\phi(t)},\quad a_{bi}\,a_{bi} = a^2,\qquad(11.11c)$$

with all amplitudes and phases real. Inserting (11.11) in (11.8) we easily obtain the relations:

$$\cos^2\theta\dot{\nu} = gK,\qquad(11.12a)$$

$$\sin^2\theta\dot{\delta} = gK,\qquad(11.12b)$$

$$a^2\,\dot{\phi}\left(1 - \frac{\dot{\phi}}{2}\right) = gK,\qquad(11.12c)$$

$$\dot{\nu} - \dot{\delta} + \dot{\phi} = 0,\qquad(11.12d)$$

with $K = a\sin\theta\cos\theta$; to which we must add, as usual, the "momentum conservation" equation

$$\cos^2\theta - a^2(1 - \dot{\phi}) = \frac{\delta\pi}{N_{CD}}.\qquad(11.13)$$

Note that, as seen already in Chapter 4, the rhs of (11.13) is determined by the size of the "zero-point" quantum fluctuations of both the nuclear matter and the π-field.

Solving for the phase velocities $\dot{\nu}$, $\dot{\delta}$, $\dot{\phi}$, the angle θ and the amplitude a^2, for the energy gain per N/Δ one obtains in a straightforward way

$$\frac{\Delta E}{A} = -\omega_{\bar{q}q} g K \left[\frac{4 - 3\dot{\phi}}{2 - \dot{\phi}} \right], \tag{11.14}$$

independent of A. Minimizing the energy one finally gets for the N-density

$$\cos^2 \theta = 0.82, \tag{11.15a}$$

and, complementarily, for the Δ-density:

$$\sin^2 \theta = 0.18, \tag{11.15b}$$

while for the π-amplitude

$$a^2 = 0.63, \tag{11.15c}$$

and for the energy gain per nucleon

$$\frac{\Delta E}{A} \cong -60 \text{ MeV}, \tag{11.16}$$

independent of A. This latter feature of the "central potential" that we have just obtained is quite fundamental and is usually referred to as the "saturation" property of the nuclear interaction. In fact, if the binding energy were due to the summation over the nucleon pair-interaction energies, one would expect it to increase like $\frac{A(A-1)}{2}$, and not like A, as experimentally observed and theoretically derived in Eq. (11.16). However this latter result does not exhaust the simple predictions that can be obtained from the Euler-Lagrange equations: by restoring their $\vec{x}$-dependence we can easily demonstrate that the nucleon field amplitude $n_{st}(\vec{x})$ obeys the following Schrödinger equation

$$i \, \dot{n}(\vec{x}, t) = -\frac{\vec{\nabla}^2}{2m_N} n(\vec{x}, t) + \left[V_0(\vec{x}) + V_{SO}(\vec{x})(\vec{L}.\vec{S}) \right] n(\vec{x}, t) \tag{11.17}$$

with the "spin-orbit" potential

$$V_{SO}(\vec{x}) = -a_{SO}^2 \frac{1}{r} \frac{dV_0}{dr}, \tag{11.18a}$$

and

$$a_{SO} \simeq 1 \ F, \tag{11.18b}$$

as required by the SM to explain, for instance, the existence of magic numbers. Note that in Hypernuclei — the nuclei in which one or more nucleons are replaced by Λ or Σ hyperons — the spin-orbit potential is one order of magnitude smaller, which shows that the dynamical mechanism that confines Λ's or Σ's inside a hypernucleus is quite different from that which keeps nucleons together in a nucleus. We will come back to this interesting question in Section (11.4).

Starting from (11.17) one can now work out the physics of the SM, and of other nuclear models (such as the droplet model) that are found to account for important aspects of nuclear phenomenology. This being, however, outside the scope of this book, we shall now turn to some recent problems where the conventional (and scarcely motivated) models have conspicuously failed.

11.3 The EMC Effect and the Coulomb Sum Rule

One of the striking new nuclear phenomena that the high-energy accelerators have uncovered is what has been called the EMC Effect. The European Muon Collaboration (EMC) has carried out over a ten year period experiments at the CERN-SPS (Super Proto Synchrotron) in the field of deep-inelastic μ-scattering off a proton or a nuclear target. The interest of experiments of this kind lies in the fact that (as shown for the first time in the experiments at the Stanford Linear Accelerator Center (SLAC) at the end of the Sixties) when probed by the highly virtual photons produced by the scattering of leptons (electrons or muons) at large momentum transfer, hadrons such as the nucleons exhibit a very simple structure, that can be interpreted in terms of them being composed of Quark/Partons, pointlike charges that scatter **incoherently** the virtual radiation produced by the inelastic lepton scattering.

The experimental observations are all in agreement with the picture of the process reported in Fig. 11.1, where it is assumed that the lepton-nucleon scattering involves incoherently, in a Rutherford-like way, the pointike Quark/Parton constituents, thus leading to a cross-section

$$d\sigma^{\gamma^* N} = F_2(x_B) \otimes d\sigma^{\gamma^* q}, \tag{11.19}$$

obtained by covoluting a Quark/Parton distribution $F_2(x_B)$, function of the Bjorken variable $x_B = \frac{Q^2}{2M\nu}$, with the pointlike (Mott) γ^*-Quark cross-section $d\sigma^{\gamma^* q}$. The meaning of x_B is quite simple, it denotes the fraction of the nucleon 4-momentum P_μ carried by the scattering Quark-Parton.

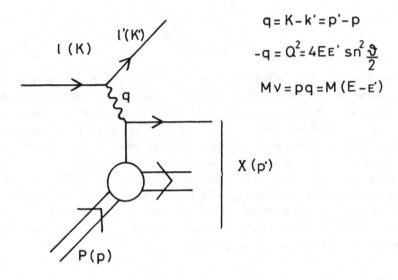

Fig. 11.1. The Feynman diagram of lepton-nucleon deep inelastic scattering, and its kinematics.

Eq. (11.19) provides an extremely accurate description of deep-inelastic lepton nucleon scattering data, a fact which appears all the more surprising if we recognize the full implications of the Rutherfordian picture engendered by Fig. 11.1. Indeed, according to such picture the struck Quark/Parton should be ejected as a **free** particle in the final state configuration, leaving behind the nucleon debris, whose quantum numbers coincide with those of a two-quark state. We now know very well that no free quark has ever been seen in an experiment of this type (or for that matter in any other experiment involving hadrons), so that *stricto sensu* the picture in Fig. 11.1 is physically meaningless, requiring more than a nominalistic stratagem — the mysterious notion of "Quark Confinement" — to recuperate some meaning. This is certainly not the place to engage in a discussion that is central to QCD and modern high-energy physics, thus from the above discussion it suffices to retain the "message" that deep-inelastic lepton-nucleon scattering does resolve, inside hadrons, **pointlike charge structures** (the Quark/Partons) which contribute **incoherently** to the scattering cross-section.

Let us now turn our attention to a large nucleus. What do we expect? Before the EMC experiments — at the beginning of the Eighties — the answer appeared straighforward. In fact, if one considered the ratio $R(x_B) = \frac{F_2^A(x_B)}{F_2^D(x_B)}$, between the structure-functions (Quark/Parton distributions) $F_2(x)$ for a nucleus of atomic number A and the loosely bound deuteron D, this ratio is obviously expected to equal

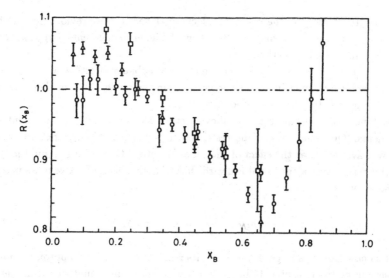

Fig. 11.2. $R(x_B)$ versus x_B in iron, observed at SLAC ($\bigcirc$), BCDMS ($\triangle$) and EMC ($\square$).

unity; for if the Quark/Parton scattering is incoherent, "a fortiori" no coherence is to be expected among nucleons, whose binding energies are much smaller than the energy transfers involved in the deep-inelastic process. The big surprise — the EMC effect[2] — that came out of the experiments can be appreciated by looking at the data reported in Fig. 11.2: a very peculiar deviation from the expected unity is observed, having its minimum (at about 0.8) for $x_B \simeq 2/3$, and exceeding 1 by some 5% for small values of x_B and large values of Q^2. What are we to make of this strange behaviour? Judging from the large number of unsuccessful analyses that have been carried out ever since, the EMC effect **does** pose a very big problem to the generally accepted view of NP. And the reason is obvious: the large deviations of $R(x_B)$ from unity simply mean that even at the extremely small space-time distances probed by the deep-inelastic virtual photons the nucleus is **not** a collection of independent, incoherent nucleons. The Quark/Partons that sit inside the nucleons have **some** way to know that their parent nucleons are part of a larger object, whose peculiar dynamics manages to distort distinctively, and universally, their momentum distributions.

From what we have been learning throughout this book the (likely) explanation of the EMC effect is that the nuclear interaction, whose two-body, short range forces can never interfere with the deep-inelastic dynamics (due to the large disproportion between the NP space-time scales (10^{-13} cm, 10^{-22} sec) and the deep-inelastic space-time scales

(10^{-15} cm, 10^{-25} sec), by the coherent effects analyzed in the previous Section is in fact promoted to a rôle that may stands the much higher, incoherent energy exchanges of the deep inelastic interaction.

We are here in presence of a kind of "nuclear Mössbauer effect" where, due to the highly coherent dynamics involving the nucleons in the nucleus, the nucleus "recoils" coherently against the deep inelastic virtual γ^*.

I shall now demonstrate that the above picture does lead to good agreement with experiments. Our analysis shall be restricted to the region of large x_B.[c] In Section (11.2) we have seen that the effect of the coherent interaction is (neglecting the small spin-orbit term) equivalent to a downward shift of the nucleon effective mass m_{eff} by some 60 MeV, thus we have

$$m_{eff} = m_N - \frac{\Delta E}{A} \cong m_N - 60 \text{ MeV.} \qquad (11.20)$$

Let us now look at a large A nucleus in its rest frame; to a good approximation its nucleons can be thought of as filling up the levels of a Fermi-liquid with 4-momenta

$$p_\mu = (E, \vec{p}) \qquad (11.21)$$

where $\vec{p}$ is the spatial momentum ($|\vec{p}| \leq p_F$, p_F is the Fermi-momentum) and $E = \sqrt{\vec{p}^2 + m_{eff}^2}$ its relativistic energy. When a nucleon filling up such levels is scattered by a deep-inelastic γ^* the x_B of its "scattered" Quark-Partons will be

$$x_B^A = \frac{Q^2}{2(p \cdot q)}, \qquad (11.22)$$

instead of the rest frame $x_B = \frac{Q^2}{2M\nu}$. Thus in the nuclear case the relevant Quark/Parton distribution function is in the variable x_B^A, which can be further written as (θ is the polar angle of the vector $\vec{p}$)

$$x_B^A = x_B \frac{m}{m_{eff}} \left[1 + \frac{p^2}{2m_{eff}^2} + \frac{p}{m_{eff}} \cos \theta \right]. \qquad (11.23)$$

Note that averaging over the Fermi-sphere ($|\vec{p}| < p_F$) $\langle x_B^A \rangle > x_B$, which qualitatively explains why R(x_B) is in the large x_B-region smaller than 1. This is the region in fact where the single nucleon $F_2(x)$ decreases so fast that, shifting the effective

---

[c]A full treatment of the EMC effect, including the interesting very small x_B-region ($x_B < 0.01$), can be found in Ref. [3].

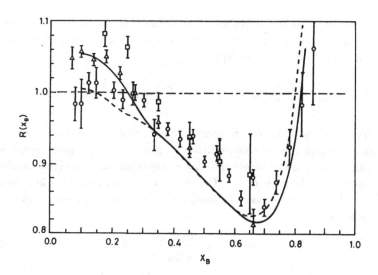

Fig. 11.3. The data of Fig. 11.2 compared with the prediction (11.24) for different values of Q^2 (14–200 GeV^2 and 2–10 GeV^2 for EMC data and SLAC data respectively).

x_B ($\left\langle x_B^A \right\rangle$) to values higher than x_B, leads to the suppression that is experimentally observed. Quantitatively the structure function $F_2^A(x_B)$ is given by a simple convolution

$$F_2^A(x_B) = \frac{3}{p_F^3} \int_0^{p_F} p^2 dp \int_{-1}^{+1} d\cos\theta \int_0^1 d\xi \, F_2^D(\xi)\delta(\xi - x_B^A), \qquad (11.24)$$

which is confronted with the experimental data in Fig. 11.3. Due to the simplicity of the calculation the agreement is really remarkable.

It appears both fortunate and fascinating that deep-inelastic scattering has given us a way to probe the "rigidity" of the coherent nucleons' motion inside a nucleus. The question now is: is there a way to test the fundamental idea that the nucleons inside a nucleus are a coherent mixture of a nucleon and a $\Delta\pi$-state, as predicted by our solution of the CE's of the $N\Delta\pi$ interaction (Eq. (11.15))? In fact according to such theory which the wave-function of the "dressed" nucleon of a nucleus is

$$| N \rangle_{Nucleus} = \cos\theta \, | N \rangle + \sin\theta \, | \Delta\pi \rangle? \qquad (11.25)$$

Fortunately the answer is positive, we only need analyse the quasi-elastic process through which a virtual photon γ^* of energy ω and 3-momentum $\vec{q}$ (produced by a lepton beam) scatters a single nucleon off a given nucleus of charge Z and neutron number

N (A=Z+N). Based on purely kinematical arguments, it can be shown that the double differential quasi-elastic scattering cross section can be written ($Q^2 = \vec{q}^2 - \omega^2$; Ω is the solid angle of the scattered lepton)

$$\frac{d\sigma}{d\Omega d\omega} = \sigma_{Mott}\left\{\left(\frac{Q^2}{\vec{q}^2}\right)R_L(\vec{q},\omega) + \left[\frac{1}{2}\left(\frac{Q^2}{\vec{q}^2}\right) + tg^2\theta/2\right]R_T(\vec{q},\omega)\right\}, \qquad (11.26)$$

where $R_T(\vec{q},\omega)$ is the transverse response function, that receives contributions from meson exchange and Δ excitation, which both grow with the lepton energy loss ω. The longitudinal response function $R_L(\vec{q},\omega)$, on the other hand, receives contributions from the protons of the nucleus only. One may show in fact that if the nucleus is well described by a dense Fermi-gas of nucleons (N neutrons and Z protons) the Coulomb Sum Rule holds:

$$C(\vec{q}) = \int d\omega \frac{R_L(\vec{q},\omega)}{[G_E(Q^2)]^2} = Z, \qquad (11.27)$$

where $G_E(Q^2)$ is the charge form-factor of the proton. The meaning of (11.27) is quite straightforward: if one integrates the longitudinal response function over the virtual photon energy and divides by the square of the proton charge-form factor $G_E(Q^2)$, thus taking into account the extended nature of the charge distribution of the proton, the result should just be the number of protons probed by the virtual photon. Measurement of $C(\vec{q})$ in ^{56}Fe yields[4] for $|\vec{q}| = 1.14\,\text{GeV}$

$$C(\vec{q}) = 0.76 \pm 0.23, \qquad (11.28)$$

and the $R_L(\omega)$ reported in Fig. 11.4. The substantial deviation from one of $C(\vec{q})$ (Eq. (11.28)) strongly suggests that the nucleus is not just a dense Fermi gas of nucleons, and this agrees very well with our finding that the "bare" nucleon content of the nucleus is $\cos^2\theta = .82$ smaller than in the conventional NP picture. This fraction should be compared with the experimental value (11.28), the reason being that $\sin^2\theta$ portion of the "dressed" nucleon, living in a $\Delta\pi$-state, cannot be **longitudinally** excited to yield a free nucleon, because the charge density operator, that is coupled to the longitudinal virtual photon, cannot induce transitions between an initial $\Delta\pi$-state and a final "bare" nucleon. Indeed the diagonal nature of the charge operator — the $\vec{q} = 0$ component of the charge density operator — forbids transitions between orthogonal states.

In this way two experimental results that have been causing grave problems to the NP community not only find a very natural explanation within our coherent picture of the Nucleus, but turn out to test in a rather unique fashion two of the main aspects of

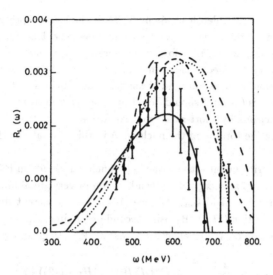

Fig. 11.4. Data on the longitudinal response function $R_L(\omega)$, compared with the present theory (continuos line), and different conventional models.

the theory presented in this Chapter: the collective motion property of the "dressed" nucleons and their "mixed" wave-function (11.25).

11.4   Λ's and Σ's inside the Nucleus

Through a "strangeness exchange" mechanism:

$$\pi^- + N \to (\Lambda, \Sigma) + K \tag{11.29}$$

modern high-energy physics has succeeded in "transmuting" one of the nucleons of a generic nucleus into a Λ or a Σ hyperon, with very little kinetic energy in such a way that the hadronic force between nucleons and hyperons is capable to bind the hyperons to the nucleus, thus forming a new object: the Hypernucleus. But what kind of force acts between the nucleons and the hyperons? Orthodox NP has no doubt: the two-body force that can be obtained from the nucleon-nucleon force by the approximate eightfold-way (SU(3)) symmetry, that since a long time is well known to tie strange particles, such as the Λ (1115) and the Σ (1190), to the nucleons. The problem is, however, that it has been experimentally ascertained that the stable configurations of a Hypernucleus are similar to those of the nucleons, i.e. their single particle levels are those of a fermion moving non-relativistically in a central potential, spanning the whole nucleus and with a depth of about 30 MeV, half the depth of the potential seen by a nucleon. Again, the question now is: how an extraneous body, like a Λ or a Σ, bound by very short range forces to nuclear matter manages to move almost freely inside the very crowded nucleus, and achieve its stable ground state in S-wave, whose wave-function is spread around throughout the nucleus. The message to the natural philosopher is (at least should be) clear: the hyperon **does not** interact with the individual nucleons by short-range nuclear forces, but is trapped in a coherent system comprising the totality of the nucleus. And this is what our theory tells us is in fact happening.

In order to describe quantitatively the hypernuclear interaction P.G. Ratcliffe and I[5] studied the interaction energy of a given hyperon of very small momentum (hence kinetic energy) in the π-field $\Phi_a(\vec{x}, t)$ generated by the coherent dynamical evolution, discussed in Section (11.2). By using second-order time-dependent perturbation theory, the interaction energy or effective potential, V_{eff}, is given by:

$$V_{eff} = -\frac{i}{4m_Y} \int_{-\infty}^{+\infty} dt \, \langle Y \mid T(H_I(t/2)H_I(-t/2)) \mid Y \rangle_{CD'} \tag{11.30}$$

where the interaction energy H_I will be explicitly given later. The expectation value is to be evaluated over a coherence domain (CD) whose size is (see Eq. (11.6)) about 4.5 F. Inserting a complete sum over intermediate states (11.30) becomes

$$V_{eff} = -\frac{1}{2m_Y} \sum_{Y^*} \int_0^\infty dt \, \langle Y \mid H_I(\vec{x}, t/2) \mid Y^* \rangle \langle Y^* \mid H_I(\vec{x}', -t/2) \mid Y \rangle, \tag{11.31}$$

where the intermediate states, Y^*, of relevance to our calculation are the spin-$\frac{3}{2}$ Σ^*, the Σ and the Λ themselves. The higher mass states such as the $\Lambda(1405)$, give negligible contributions for reasons which will become clear later. A pictorial representation of Eq. (11.30) is given in Fig. 11.5, showing that according to our theory the hyperon interacts with the nucleus through forces of the longest possible range that have their origin in its coupling to excited, virtual hyperons via the coherent π-"condensate", that we have derived in Section (11.2).

Noting that the energy denominators coming from the t-integral in (11.31) determine an effective cut-off on the momentum of the intermediate state of the order of several hundred MeV, which is much larger than the impulse corresponding to the size of the coherence domain (~ 44 MeV), the space integrals in the expectation value are restricted to the region $\vec{x} \geq \vec{x}'$. Performing the necessary algebra we arrive at the following expression for the effective potential for a hyperon inside the nuclear

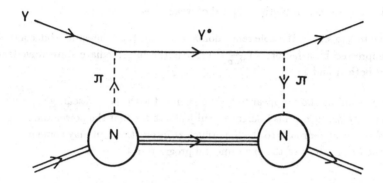

Fig. 11.5. The hyperon-nucleus interaction via the coherent pion condensate, to this diagram then should be added the contribution with the π-lines crossed.

medium:

$$V_{eff} = \frac{4}{3} G_{\pi Y Y^*} \omega_q^2 \left\langle |a_\pi(\vec{x})|^2 \left[\frac{1}{m_{Y^*} - m_Y - \omega(\vec{x})} + \frac{1}{m_{Y^*} - m_Y + \omega(\vec{x})} \right] \right\rangle_{CD}, \quad (11.32)$$

where $|a(\vec{x})|^2$ corresponds to the (space-independent) rescaled π-amplitude defined in (11.7c), $\omega(\vec{x}) \sim 90$ MeV is the renormalized pion frequency ($\omega(\vec{x}) = \omega_q(1 - \dot\phi)$, see Eq. (11.11c)) inside the coherent nuclear medium. The dimensionless coupling constant $G_{\pi Y Y^*}$ is of order unity and is defined in analogy with (11.8d) as:

$$G_{\pi Y Y^*}^2 = \frac{8\pi^2}{9} \frac{\tilde{g}^2 q^2}{4 m_Y m_Y^* \omega_q^3} \left(\frac{A}{V} \right),$$

with q and ω_q as in Eq. (11.8d), and $\tilde{g}$ is fixed by the $\Sigma^* \to \Lambda(\Sigma)\pi$ decay width. It should now be clear why the higher mass baryonic strange resonances give negligible contributions. The reasons are two-fold: i) the suppression due to the energy denominators of Eq. (11.32), and ii) in most cases the coupling (as deduced from the relevant decay widths) is very small.

Putting all the information we have ($G_{\pi Y Y^*}^2 \sim 1.2$, $\alpha^2(0) \sim 0.1$ and $\omega(0) \sim 90$ MeV) into the calculation we obtain the following results:

(i) the hypernuclear potential V_{eff} has the same shape as the nuclear potential with a depth ~ 30 MeV for both Λ and Σ (slightly shallower in the case of Σ);

(ii) the width of the states (the imaginary part of V_{eff} coming from the Σ^* width) can be calculated from Eq. (11.32) to be about 8 MeV for Σ hypernuclei, while in the case of the Λ the system $\Lambda\pi$ is well below the threshold for Σ^* decay and thus one expects a lifetime typical of weak interactions;

(iii) due to the size of the coherence domain we find that the spin orbit coupling is suppressed by a factor $\sim (\frac{R_{Y^*}}{R_{CD}})^2$ (R_{Y^*} is the intermediate state wave-length) for both Λ and Σ.

All three points above appear to be in agreement with the still scanty experimental informations existing on this fascinating subject. I hope that the power and simplicity of this theoretical approach to the hypernuclear interaction can play some interesting rôle in the future steps of the experimental program in this area.

References to Chapter 11

1. G. Preparata, *Il Nuovo Cimento* **103A** (1990) 1213.
2. J.J. Aubert *et al.*, *Phys. Lett.* **B123** (1988) 275;
   M. Arneod *et al.*, *Nucl. Phys.* **B333** (1990) 1;
   R.G. Arnold *et al.*, *Phys. Rev. Lett.* **52** (1984) 727;
   A.C. Benvenuti *et al.*, *Phys. Lett.* **B189** (1987) 483.
3. G. Preparata and P.G. Ratcliffe, *Phys. Lett.* **B276** (1992) 219.
4. Z.E. Meziani *et al.*, *Phys. Rev. Lett.* **52** (1985) 2130; ibid. **54** (1985) 1233.
5. G. Preparata and P.G. Ratcliffe, *Il Nuovo Cimento* **106A** (1993) 685.